Probiotics for Sustainable Food and Medicine

PROBIOTICS FOR SUSTAINABLE FOOD AND MEDICINE

Edited by:

Laghulkar Anita Marotirao

Delve PUBLISHING

www.delvepublishing.com

Probiotics for Sustainable Food and Medicine

Laghulkar Anita Marotirao

Delve Publishing
224 Shoreacres Road
Burlington, ON L7L 2H2
Canada
www.delvepublishing.com
Email: orders@arclereducation.com

ISBN: 978-1-77469-397-1 (Hardcover)

ABOUT THE EDITOR

Miss Anita Laghulkar is presently serving as Assistant professor, Department of Food Chemistry and Nutrition, College of Food Technology Naigaon (Bz) Nanded, Maharashtra affiliated by Vasantrao Naik Marathwada Krishi Vidyapith, Parbhani. She obtained her B.Tech (Food Technology) in 2016 from College of Food Technology Naigaon & M.Tech (Food process Technology) in 2018 from Maharashtra Institute of Technology Aurangabad, Maharashtra. She started her career as Assistant professor in 2018. Her M.Tech specialization in reduction of acrylamide compound from processed Food. She has published 2 research paper in international journal. She has participated in many National and international symposiums and state seminar.

TABLE OF CONTENTS

LIST OF FIGURES

ABSTRACT

Probiotics are non-pathogenic organisms (yeast or bacteria) in food that can positively affect the host's health. Probiotics have become increasingly popular in the past two decades; with the application of various products, more and more scientific evidence has shown that probiotics are beneficial to human health, and the food industry is very active in researching and promoting them. This market has integrated probiotics into different products, primarily fermented dairy products. Given this continuing trend, despite substantial scientific evidence linking these microbes to different health benefits, more research is required to identify them and evaluate their safety and nutritional aspects. This volume aims to assess the current literature on the concept and potential beneficial elements of probiotics in the literature, focusing on probiotics in food and medicine.

PREFACE

Gibson and Roberfroid launched the word "prebiotic" in 1995 to define dietary supplements that cannot be digested by the host but can have a beneficial effect through the activity of microorganisms in the gut. In most countries, probiotics account for the majority of their products. Beneficial effects of probiotic intake have been reported to include improved gut health, relief of lactose intolerance symptoms, and reduced risk of various other diseases, and some well-defined strains of *Lactobacillus* and *Bifidobacterium* are available for human use. Despite promising evidence, the effects of probiotics on human health and the safety of their use require further research, as the properties required for their functioning in the gut are currently incomplete.

Prebiotics are neither hydrolyzed nor absorbed in the gastrointestinal tract but can be used as probiotic substrates. Currently, the most commonly used is indigestible oligofructose. For convenience, the combination of probiotics and prebiotics is referred to by some authors as conbiotics and others as symbiotics. Although prebiotics appears to play a role in promoting health, this needs to be supported by further research. According to this volume, the term functional food has also been developed in recent years to describe foods that contain ingredients that have beneficial effects on host health past their nutritional value. These include products containing bioactive ingredients that improve health, such as B. Probiotics.

Given their definition, the number of microbial species that can exert probiotic effects is impressive. From a nutritional point of view, only the strains classified as lactic acid bacteria are essential, the most important of which are the genera *Lactococcus* and *Bifidobacterium*. Lactic acid microbes are Gram-positive, catalase-negative bacteria capable of producing lactic acid as the primary end product of carbohydrate fermentation. Therefore, *Bifidobacterium* is phylogenetically more traditionally listed as one of these because they utilize separate metabolic pathways from *Lactococcus lactis*, the two most important commercial lactic acid bacteria.

The book states that it is crucial to mention that since the probiotic activity is strain-dependent, strain identification is recommended to determine their suitability and output for industrial applications. This is achieved by combining phenotypic testing and genetic characterization using molecular techniques such as DNA/DNA hybridization, 16SRNA sequencing, etc. A potential probiotic strain to exhibit its beneficial effects is expected to exhibit specific desirable properties. It has been determined by *in vitro* tests that

a. Acid and bile tolerance appear to be critical for oral administration
b. Adhesion to mucosal and epithelial surfaces is a significant property for successful immune modulation, pathogen competitive exclusion, and prevention of pathogen adhesion and colonization,
c. Antibacterial action against pathogenic bacteria
d. Bile salt hydrolase action.

However, since no specific parameters are essential for all probiotic applications, the best way to characterize a strain is through target population and target physiological function-specific research.

CHAPTER 1

COMMON TYPES OF PROBIOTICS

CONTENTS

Probiotics are "healthy" bacteria that are advantageous to the health and function of the human body. Numerous probiotics are connected to the microorganisms that the body naturally generates. Consumption of fermented foods, such as yogurt and other similar items, may lead to the absorption of some of these helpful bacteria.

1.1 DIFFERENT CHEESES

Since the early 1900s, the notion of consuming more beneficial microorganisms to promote health has been researched. The notion of probiotics is very recent. It is a mix of the Greek and Latin words for life and for. The outcomes of studies on the health effects of probiotics are favorable. However, the advantages of probiotics for healthy persons or the therapy of particular ailments are not proved by inquiry. Probiotics are currently accessible in several meals and dietary supplements (Zhang *et al.*, 2016). Some experts are anxious, citing the National Center for Complementary and Integrative Health Reliable Source, that the quick expansion in probiotic usage and marketing may have surpassed the field's grasp of their advantages.

1.2 WHAT MAKES PROBIOTICS ADVANTAGEOUS?

According to Harvard Medical School, the intestines are home to 100 trillion varieties of bacteria. Each of the bacteria in the body is unique, akin to a fingerprint. Occasionally, sickness or stress can disturb the bacterial balance, producing digestive and other diseases.

Probiotics alter the makeup of the gut bacteria or the metabolic activity of existing bacteria to accomplish their desired effects. Healthy bacteria outnumber harmful pathogens in the intestines. This blocks the undesirable germs from multiplying and creating an illness or inflammation.

For example, an overpopulation of yeast in the body can lead to a yeast infection, but a healthy gut biome keeps yeast levels low. Probiotics may aid in digestion and nutrient absorption from the diet. Beneficial microorganisms can develop inhibitory or even lethal proteins or enzymes. Certain probiotic species can also enhance the immune system (Yao *et al.*, 2020).

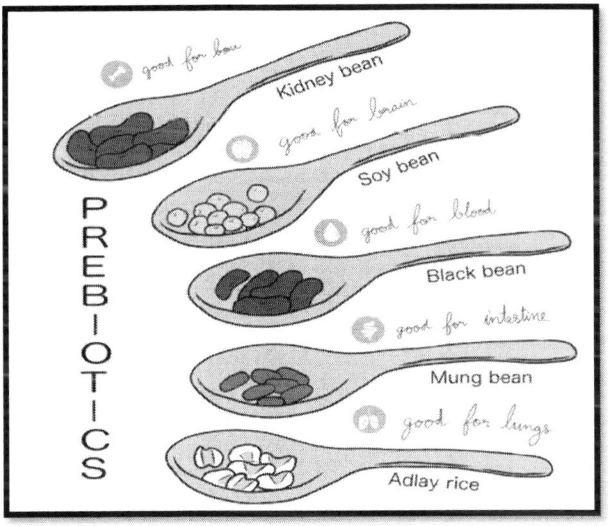

Figure 1.1: Prebiotics of bean.

Certain microbes are needed to create hormones or absorb vitamins and minerals (such as vitamin K). Childhood type 1 diabetes may be induced by inflammation produced by an insufficient gut bacterial ecology. According to American College of Gastroenterology studies, specific probiotic species may:

- help with digestion
- stop diarrhea
- cure genital infections
- stop autoimmune diseases
- improve skin problems
- prevent urinary tract infections

It is crucial, though, that not everyone will benefit from every probiotic similarly. To address a specific condition, some physicians recommend mixing probiotic pills with antibiotics (Abuajah *et al.*, 2015). As a result, both healthy and hazardous germs could be eliminated, which might create diarrhea. Probiotics help in maintaining or restoring a healthy microbiome in the gastrointestinal system.

Figure 1.2: Probiotics Treatment Microflora icons.

1.3 FREQUENTLY USED PROBIOTICS

Microbes from two basic genera make up the bulk of probiotics ingested. Genera from which probiotics of particular interest have originated include the following:

Meals and dietary supplements typically contain the bacterium *Bifidobacteria*. Considered by some to be

- Increase the immune system's capacity

The development of dangerous germs in the digestive tract is inhibited.

Help the body convert lactose into nutrients that it can utilize

Lactase, an enzyme that breaks down lactose, is generated by *Lactobacillus* bacteria. These bacteria also generate lactic acid. When exposed to lactic acid, it is simpler to manage dangerous germs(Aggarwal *et al.*, 2013). In addition, it assists in the body's absorption of minerals and serves as a source of muscle energy. The following foods include naturally occurring lactobacillus bacteria:

It is vital to understand the word "mouth".

- The intestines are quite tiny.

Infertility

Figure 1.3: Gut bacteria.

1.4 TYPICAL SPECIES OF PROBIOTICS

There are many different forms of probiotics. Distinct probiotic species have different impacts on the human body. Probiotic species names and genus names will appear on food and supplement labels.

There are several cases when the names of the genera *Bifidobacteria* and *Lactobacillus* are abbreviated to just B. or L. and linked with the names of chosen species, such as acidophilus. As a consequence, *Lactobacillus acidophilus* probiotic is obtained. This is how the name will appear on supplements and consumables (Aponte et al., 2020).

On food and supplement labels, these six probiotic strains are discovered:

B. animalis: Dannon's Activia yogurt contains this particular species of *B. animalis*. In addition to improving digestion, it protects against food-borne pathogens. Additionally, it can benefit general health by strengthening the immune system.

Vagina and the gastrointestinal tracts are both home to *B. breve*. Both locations benefit from the antifungal and antibacterial qualities of this medication. Carbohydrate digestion improves the absorption of nutrients by the body. In addition, it breaks down plant fibers so that the human body may absorb them.

In raw milk, *B. lactis* is a bacterium. Probiotic infant formula from Nestle contains it as an ingredient. Additionally, it acts as a Launchpad for other apps. Tofu, yogurt, curd cheese, and cheeses to pair with the wine.

B. longum is a kind of bacterium that resides in the gastrointestinal tract.

It contains antioxidant qualities and helps break down glucose.

Vagina and the small intestine are abundant in *Lactobacillus acidophilus* (*L. acidophilus*). Yogurt and fermented soy products like miso include probiotics (Wescombe *et al.*, 2012).

Lactobacillus reuteri is a species that may be discovered in the mouth and the stomach. One research indicated that it decreased the bacteria in the mouth that causes tooth decay. The digestive system may also benefit from it.

1.5 THE USE OF PROBIOTICS SHOULD BE INCREASED

Is it possible to discern if probiotics are good? Consult a doctor to discover if they are safe for to take. Probiotics are typically considered safe to ingest in case of a weak immune system or a significant medical condition. Inquire with a doctor or pharmacist about the dose, too.

Begin taking probiotics in the diet by introducing a few items that naturally contain probiotics. Probiotics can have favor the digestive system, so maintain track of how the health improves over time. A few quick and easy meal ideas are Yogurt from Greece, for example, alcoholic drinks such as bourbon and rye whiskey, gruyere cheese, Cheddar cheese, and Grated Parmesan.

Vegetables fermented such as:

- The sour cabbage
- Cucumbers in vinegar

There are various commercially available probiotic dietary supplement alternatives.

A mixture of probiotics, rather than a single one, has been demonstrated to be more effective. Colony-forming units, or CFUs, measure the number of bacteria needed to build a colony. Probiotics are not subject to FDA regulation (FDA) because they are considered dietary supplements. As a result, neither the Food and Drug Administration nor probiotic businesses are required to present proof to back their effectiveness claims (Valero-Cases *et al.*, 2020).

One of these bacteria's vital advantages is its capacity to create antibiotics, such as lactolin, that are aimed precisely at dangerous microbes. Lysine, a critical amino acid, is also generated by this probiotic.

Lactobacillus plantarum, the body's natural probiotic, is only present briefly. Moreover, it is built to endure a lifetime. This chemical can be tolerated from vitamins in the mouth to intestines and colon waste. Due to the optimal temperature for growing, *Lactobacillus plantarum* is so similar to that of the human body that this supplement is designed to be taken internally. In the fight against harmful microorganisms, *Lactobacillus plantarum* is a great guest probiotic. Removing the bad guys, enhances the natural microbes and increases our resistance to illness.

Lactobacillus plantarum has been taken by humans for all of recorded history because it is a great immune system booster. Foods that contain this strain include:

- Sourdough bread cooked
- Fermented cabbage

1.6 PROBIOTICS MYTHS AND MISCONCEPTIONS

There is much information to wade through regarding the numerous probiotics on the market. As a result, some widespread myths and misunderstandings concerning these products exist. Here are some of the most typically misinterpreted probiotic facts:

There are many different forms of probiotics. Probiotic strains can have a wide range of effects. The advantages of probiotics depend on the strain, which must be considered. If probiotics are taken from the same genus or species, they may not all have the same advantages. The strain's published scientific findings must be connected to the probiotic strain's scientific discoveries (Tadesse, 2012).

A strain of bacteria is like a dog's breed; while all dogs are members of the same genus and species, they all have distinct physical qualities and talents.

There is no certainty that more is better: Calculating probiotics in a meal is commonly done by calculating the quantity of CFUs, or live bacteria. Clinical research and the quantity that has been useful for various strains and the health impact under consideration define the appropriate amount of probiotics. More is not necessarily better in cannabis; certain strains function better at varying amounts. There is no such thing as a safe or harmful number of germs to consume to achieve the appropriate outcomes. Probiotic dosage recommendations vary depending on the strain and intended impact. A multi-strain probiotic product is likely to outperform a single-strain probiotic product. Once again, whether or not a particular strain or a combination of probiotic strain has been proven useful for a specific health benefit depends on the stated objective.

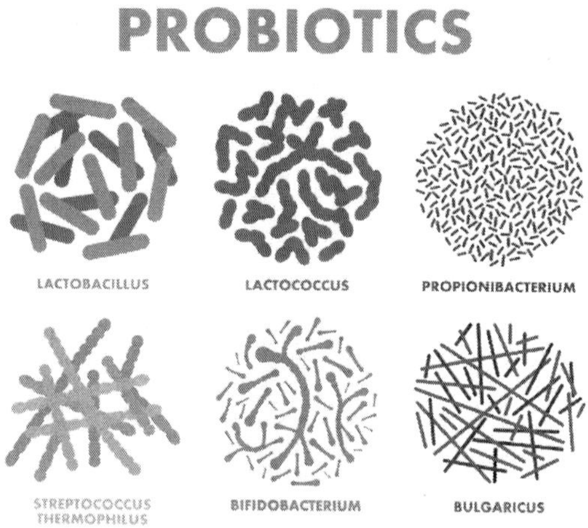

Figure 1.4. Probiotic Bacteria.

Probiotic strains are identified by alphanumeric identification, genus, species, and other distinguishing characteristics. The number of colony-forming units (CFUs) needed varies substantially depending on the bacterial strain. Higher CFU counts may not always indicate higher performance.

The amount stated on the label must be supported by scientific evidence that it positively impacts human health. In multi-strain probiotics, there should be a strong interaction between the strains. IBS is a severe and increasingly prevalent intestinal disorder. However, recurrent farts may imply more grave medical conditions. Plantarum can minimize flatulence (much to the dismay of Swedish comedians). He offered a rose hip and *Lactobacillus plantarum* beverage to sixty patients as a placebo or to test his hypothesis (Smolyansky, 2010). Significantly fewer participants utilizing *Lactobacillus plantarum* suffered flatulence than those getting a placebo. The effects were investigated for a year to establish their durability. Even one year after supplementation ceased, the *Lactobacillus plantarum* group maintained lower levels of flatulence and improved gastrointestinal function, demonstrating that the favorable advantages of the supplements were not ephemeral.

The complete list of digestive benefits is as follows:

- Drastically decreases IBS symptoms
- Impaired intestinal barrier function
- Battles varied intestinal illnesses (Eg. *Clostridium difficile*)
- Decreases bloating

Numerous species make up one of the largest and most popular probiotic genera. *Lactobacillus* is a bacterium found in fermented foods, including yogurt, cheese, sourdough, and in the human digestive system. Diarrhea treatment and prevention are among the health advantages of *Lactobacillus* that are most widely highlighted. It has been demonstrated that *Lactobacillus* is effective in treating and preventing bacterial and infectious diarrhea. In addition to alleviating IBS symptoms, *Lactobacillus* can also treat other common digestive problems and enhance overall gut health overall (Stavropoulou & Bezirtzoglou, 2020). *Lactobacillus* is also extensively used to treat stomach discomfort, inflammation of the colon, baby colic, and constipation. How probiotic strain names are selected

Lactis is the species, BI-07 is the strain, and *Bifidobacterium* is the genus when discussing Bifidobacterium lactis BI-07. However, what is its significance?

- Genus: The genus is the most comprehensive classification scheme. It encompasses many bacterial species that belong to the same broad group yet have various features and health advantages.
- Species: This term defines a particular species. Despite their similarities, there are a few slight differences between the traits of each bacterial strain within a species.
- Strain: A distinct strain of bacteria is one form of bacterial classification. Each strain of bacteria present in the body performs the same job.

1.7 WHAT DIVIDES *LACTOBACILLUS* FROM *BIFIDOBACTERIUM*?

Lactobacillus and *Bifidobacteria* differ substantially in the beneficial compounds they manufacture.

Lactobacillus ferments refined carbohydrates and makes lactic acid, whereas *Bifidobacteria* usually produce short-chain fatty acids (SCFAs) and lactic acid (Arora & Baldi, 2015).

Another distinction is where they reside (or grow). The small and large intestines are the natural habitats of *Lactobacillus* and *Bifidobacterium*, respectively.

As it takes longer for *Bifidobacterium* strains to reach the large intestine, when a probiotic supplement is taken, Lactobacillus bacteria travel to and grow in the small intestine first.

1.7.1 Bifidobacterium

Bifidobacteria, which are Gram-positive anaerobic bacteria, is needed for normal intestinal inflammation and aid in the digestion of carbohydrates.

Bifidobacteria is one of the most prevalent genera in the gut microbiome of the majority of humans, comprising 3 to 6 % of the entire microbiome.

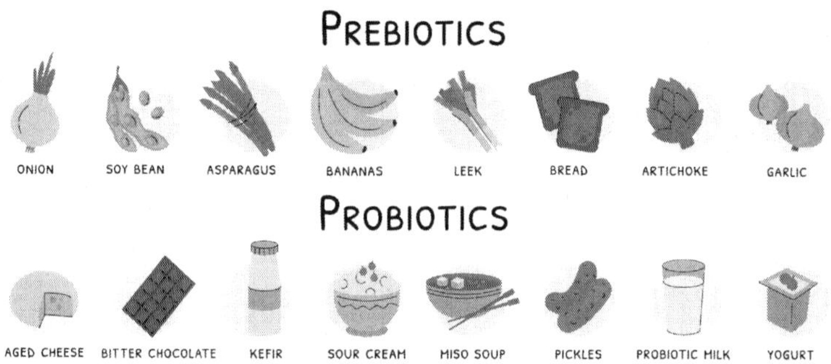

Figure 1.5: Prebiotic food products.

There are around 48 varieties of *Bifidobacterium*, each of which has a particular effect on health. *Bifidobacteria* provides many health advantages, some of which are detailed below.

- A robust immune system
- A physiological reaction to stimuli that is more effective
- Improved digestion and relief from gas and constipation*
- Mood management

1.7.2 Lactobacillus

Lactobacillus, a Gram-positive, non-spore-forming bacterium, is possibly the most well-known probiotic despite accounting for less than 0.01 % of the total quantity of bacteria in the gut. Although this seems to bc a modest quantity, it is not negligible (Saraf *et al*., 2010).

According to studies, Lactobacillus can support the gut barrier function by eliminating potentially dangerous microorganisms.

There are over 170 species of Lactobacillus, and they give the following general health benefits:

- Facilitates lactose and milk sugar digestion
- Digestive health and regular bowel movements
- Has a Favorable impact on total cholesterol levels
- Enhances immune system function
- Promotes better sleep

Lactobacillus ferments refined carbohydrates to make lactic acid, whereas *Bifidobacteria* produce predominantly short-chain fatty acids. The colon and small intestine generally contain *Lactobacillus* and *Bifidobacterium*, respectively. *Lactobacillus* is connected with a healthy digestive system, regular bowel movements, and improved sleep, while *Bifidobacteria* is associated with a healthy immune system and management of mood.

1.8 STANDARD PROBIOTIC ORGANISMS AND STRAINS

Indeed, the genera *Bifidobacteria* and *Lactobacillus* contain multiple species and strains.

Although there are around 8,000 individual probiotic strains, scientists have not yet discovered them all (and probably never will).

Some bacterial strains may never be comprehended, but others have a substantial history of study and medicinal usage. These are described below:

It has been demonstrated that *Bifidobacterium lactis* HN019 promotes regularity and lowers constipation.

By lengthening colon transit time, *Bifidus regularis*, also known as *Bifidobacterium animalis* DN-173 010, has increased regularity and alleviated constipation (Attri et al., 2021).

Bifidobacterium longum BB536: maintains bacterial balance and homeostasis in the gut, which is favorable to the immune system and alleviates digestive diseases.

Bifidobacterium lactis UABla-12 and *Lactobacillus acidophilus* DDS-1 can alleviate bloating and pain in adults with irritable bowel syndrome.

Lactobacillus casei Shirota and *Bifidobacterium lactis* DN-173 010 may boost stool frequency and consistency to cure constipation and support regular bowel function.

Lactobacillus fermentum RC-14 and *Lactobacillus rhamnosus* GR-1 have been demonstrated to promote vaginal health and lessen the risk of bacterial and yeast infections.

Lactobacillus rhamnosus JB-1 positively regulates the neurotransmitter GABA in the brain, which may alleviate stress-related symptoms.

Lactobacillus casei BF1, BF2, and BF3 can potentially reduce *Helicobacter pylori* and *Staphylococcus aureus*; two dangerous bacteria are known to infect people and create health concerns (Sistek et al., 2006).

Lactobacillus reuteri NCIMB 11951, *Lactobacillus reuteri* NCIMB 701089, and *Lactobacillus acidophilus* ATCC 314 are three lactobacilli strains that support cardiovascular health and normal cholesterol levels.

Lactobacillus plantarum DSM 9843 may lower gas and flatulence in IBS patients.

1.9 WHY ARE PROBIOTICS VALUABLE?

After dissecting and examining the various significant strains that may be incorporated in probiotic supplements, it is obvious why they are valuable for maintaining overall health. They give a lot of added health advantages. Prebiotics are beneficial components of probiotic supplements that support and promote health in almost all body systems.

1. A healthy mix of "good" and "bad" bacteria in the stomach improves digestion and absorption of nutrients.
2. Feeding "good" bacteria that may be overwhelmed by oxidative stress.
3. Promotes proper peristalsis transforms food into nutrients that the body can use for energy, cell development, and repair (Sidhu *et al.*, 2015).

4. Promotes mineral absorption from the meals, particularly those that support bone health, such as iron, calcium, magnesium, and zinc.
5. Assists the body's infrequent metabolism of common toxins.
6. Supports immune cell activity and helps the body create natural antibodies for healthy immunological responses in response to seasonal and environmental stressors.

1.10 WHAT SITUATIONS REQUIRE THE USE OF PROBIOTICS?

Because the human body contains more bacteria than cells, sustaining an excellent bacterial balance is crucial for general health. Probiotics can help overall health daily but can also be useful at times (Singh *et al.,* 2011).

For example, the doctor may urge you to integrate probiotics into the antibiotic prescription. Probiotic pills daily can also assist in boosting digestive health and prevent indigestion, bloating, and gas. This is especially true following meals.

1.11 WHAT EFFECTS DO DIFFERENT PROBIOTIC STRAINS HAVE?

Fructo-oligosaccharides (FOS), a form of soluble fiber in various grains, fruits, and vegetables, are vital for probiotic synthesis. This type of fiber is known as prebiotic fiber. Simply put, prebiotics is nutrients that bacteria use to be alive and active in the gut.

When probiotic and prebiotic strains are mixed, the body can flourish. By promoting the growth of beneficial bacterial strains and driving away the bad ones, benefits of superior digestion can be obtained, such as increased energy, improved weight control, and making feeling good the new normal (Singh & Natraj, 2021). A high-quality probiotic can be integrated into the diet to enhance the immune system, digestive system, and general health. Complete Probiotics Platinum® by 1MD Nutrition contains 11 distinct strains and assures 51 billion live cultures each tablet for the strongest digestive and immune system support possible.

Researchers have yet to identify every single probiotic strain that exists. According to research, there may be roughly 8,000.

Summary

It is vital to purchase supplements from a trustworthy, high-quality provider. Experts recommend choosing a probiotic that is targeted to the condition to be treated, contains multiple strains, has 5 to 100 billion colony-forming units (CFUs), and has been tested by a third-party lab to ensure that it contains what it claims to and that the bacterial strains are active and carrying out their functions. Probiotic supplements supply numerous health advantages, from immune support to gastrointestinal help; however, several unique bacterial strains are available, each with specific attributes.

1.12 LACTOBACILLUS PLANTARUM

Burn infections can be avoided (topically), and *Lactobacillus plantarum* immunity benefits your immune system.

- To reduce the number of cases of colds and flu
- To reduce the duration and severity of a cold or flu.
- To accelerate burn healing
- For antifungal treatment
- To reduce the occurrence of pancreatic sepsis Due to a bacterial or viral infection
- To strengthen the immune systems of HIV-infected children
- To reduce infection risk in the weeks and months following surgery.
- To reduce the risk of pneumonia

1.12.1 Health Advantages of Lactobacillus plantarum

- Reduces inflammatory response (by increasing the production of Th2)
- Improves the quality of life for those who have experienced a variety of traumas
- Lowers the likelihood of kidney stone formation
- Reduces systolic hypertension
- Adverse Reactions to *Ligustrum plantarum*

Lactobacillus plantarum has very few side effects.

The worst-case scenario is feeling mildly nauseated.

Lactobacillus plantarum is the most common bacterial species associated with olive processing (previously *Lactiplantibacillus plantarum*). Researchers have been studying how *L. plantarum* strains process phenolics as the importance of phenolics in olive oil's health benefits has grown in recent years. When tested on olive phenolic acids, *L. plantarum* metabolized many hydroxycinnamic acids (such as caffeic, m-coumaric, o-coumaric, p-coumaric, ferulic, and sinapic acids) and some hydroxybenzoic acids (such as Gallic and protocatechuic acids) (Ayichew *et al.*, 2017).

Decarboxylase and reductase enzymes from these metabolic pathways were identified, cloned, and biochemically characterized. Through the sequential action of unidentified enzymes, *Lactobacillus plantarum* degrades the olive secoiridoid glucoside oleuropein to elenoic acid and hydroxytyrosol. The sensitivity of *Lactobacillus plantarum* to various olive-related compounds (olive oil, p-coumaric acid, gallic acid, and oleuropein) has also been investigated using a genome-scale transcriptional signature and physiological evaluation.

Lactobacillus plantarum can be found in many fermented foods, anaerobic plant waste, meat, and processed meals. This species has one of the largest genomes of the LAB and is highly adaptable and diverse. It is a Gram-positive aero tolerant bacterium that can grow at 4 % NaCl concentrations but not at 45 °C and produces both isomers of lactic acid. The bacterium *Lactobacillus plantarum* is commonly found in fermented foods such as sauerkraut and pickles, as well as Korean kimchi and Nigerian ogi. Sourdough and other fermented plant materials, as well as some cheeses and fermented sausages, are other sources of *Lactobacillus plantarum*. Due to the high concentrations of this bacteria in food, it is a likely cause of food spoilage.

Processed foods may contain slimy, souring, or greening defects. Slimy degradation is frequently restricted to the product's outer surface and may begin as solitary colonies that proliferate to form a uniform gray slime coating. It is critical to detect and count the bacteria during the various stages of vinification to assess the risk associated with the resident species (Rusu *et al.*, 2019). The importance of LAB in wine deterioration is well understood; this organism is frequently found in musts and wines due to its heterofermentative properties and is responsible for spoilage. Concurringly, acetic acid and biogenic amines may be produced.

1.13 LACTOBACILLUS BREVIS

Lactobacillus brevis, a kind of lactic acid bacterium, may be found in milk and the decomposition of various plants. Pickles and other fermented foods make use of the bacteria. They are also utilized in the production of wine and beer. *Lactobacillus brevis*, sometimes known as *Lactobacillus brevis* on labels, is found in various yogurt brands. It is also sold as a supplement for its purported health benefits. It may be present in trace levels in breast milk and the human body.

What are the advantages?

Probiotics have gained much attention in recent years. We are presently investigating how they could assist health by inhibiting the spread of pathogenic bacteria in the gastrointestinal tract. Probiotics also help to strengthen the immune system (Reid, 2016).

L. brevis may aid in healing juvenile diarrhea and increase the activity of natural killer (NK) cells in older adults. NK cells are pathogen-defeating cell that defend the body against infections and reduce the development of cancer cells.

Taking a probiotic pill or eating probiotic-rich foods may help the body fight infections and promote intestinal health. *Lactobacillus brevis* and other probiotics like it have also been linked to improved heart and gum health, and happier emotions. Antibiotics may become more effective when *Lactobacillus brevis* is present.

Lactobacillus brevis is used by winemakers to reduce the acidity of some red wines. Although it might harm taste, it can impart a subtle buttery flavor to the wine. Beer producers can use *Lactobacillus brevis* and other lactic bacteria to give their product a sour taste. These bacteria may cause the beer to deteriorate if brewers are not vigilant.

1.13.1 Surprising Benefits of Probiotics

Contradictory studies are being conducted on the safety of probiotics. The unpredictability of clinical trial outcomes is highlighted in an article published in the journal Clinical Infectious Diseases. Probiotic advocates say that the long history of safe ingestion proves that probiotics are generally safe for healthy people (Qi *et al.,* 2019). They also highlight promising findings from a few human trials, as well as animal and *in vitro* research (outside the body)

An earlier study published in Food and Chemical Toxicology in 2009 discovered that a strain of *Lactobacillus brevis* was safe for human consumption. The strain was derived from suguki, a fermented vegetable in Japan for millennia.

Human risks associated with probiotic use include:

- Systemic diseases
- Immune system concerns
- Negative gastrointestinal effects

The following foods contain *Lactobacillus brevis*:

- Yogurt from Greece
- cured cabbage
- cucumbers pickled
- bread made with sourdough

When looking for helpful probiotics, look for the terms "live and active cultures" and double-check the expiration date.

Yogurt may induce flatulence in case of lactose sensitivity since it contains milk. Instead, multivitamins can be taken. Probiotic supplements may be found at most supermarkets and health food stores. Check any supplements' expiration dates. Keep *Lactobacillus brevis* out of direct sunlight and in a cool, dry environment (Prins, 2011).

How much food can be eaten?

Adults should take 10 to 20 billion colony-forming units (CFU) of *Lactobacillus brevis* daily as a probiotic supplement. Every day, children should be administered between 5 and 10 billion CFU. These are typical probiotic supplement doses. The manufacturer will determine the exact quantity.

1.14 RISKS AND PRECAUTIONS

Probiotics are generally considered safe for healthy people to consume, according to the Mayo Clinic. Although side effects are infrequent, consult the doctor before using probiotics or any other dietary supplement (Pratap *et al.*, 2020).

The United States Food and Drug Administration (FDA) does not regulate supplements like it controls prescription pharmaceuticals. Manufacturers are not required to defend the efficacy or safety of their products.

Be aware that any supplement may include components not listed on the label. The amount of a specific component given may not always be accurate. The FDA's stance on vitamins and supplements may change in the future.

1.14.1 Will the New FDA Regulations Make Supplements Safer?

Probiotics, such as Lactobacillus *brevis*, boost immunity, improve health, and prevent illness. Probiotics may start working in the digestive system a few hours after consumption. It is more difficult to predict when long-term health gains may occur (Bowe *et al.*, 2014). There is no guarantee that taking probiotics can boost the immune system or heal digestive issues. These supplements, on the other hand, are frequently well tolerated.

Even if *Lactobacillus brevis* benefits the health, it accounts for only a tiny fraction of the overall well-being. The following aspects should also be considered to maintain health and well-being:

- eating a well-balanced diet
- physical exercise
- having adequate rest

Including probiotics in a healthy lifestyle can be considered, at least temporarily, after consulting with a doctor.

L. brevis is utilized in the production of a wide range of fermented foods all over the world. However, it can occasionally cause certain foods and beverages to spoil. *Lactobacillus brevis* is one of the bacteria responsible for most beer deterioration. Consuming meals containing probiotic strains from this species has been linked to various potential health advantages (Pandey *et al.*, 2015). This chapter discusses the organism's significant features, techniques for detecting and counting it, and its utility to customers and the food industry.

Are the most common adventitious heterofermentative *Lactobacilli* found in cheese? This group lacks the most common adventitious NSLAB in most pasteurized milk cheeses. Heterofermentative lactobacilli are seldom observed during the ripening process of cheese made from pasteurized milk in sanitary conditions. Strains are often observed in modest quantities in young (less than a month old) cheese but disappear as the cheese matures. This is because the other *Lactobacilli*, particularly *Lactobacillus paracasei* and *Lactobacillus Plantarum*, outgrow the heterofermentative lactobacilli that remain in trace proportions.

Despite the presence of NSLAB, heterofermentative lactobacilli cell density can grow in raw milk cheeses. When the conditions are favorable (for example, high temperatures and large initial numbers relative to other adventitious lactobacilli), heterofermentative lactobacilli develop to high cell densities (>10^7 CFU g1) and sustain them during cheese ripening, usually leading to a variety of problems. In most cases, however, strains enter the cheese after it has been pasteurized. For example, because these gas-forming lactobacilli infiltrate the cheese during brining, it is critical to keep the temperature low, the sodium chloride content in the brine at >16 %, and the pH at 4.5.

Slits in cheddar cheese were previously related to the formation of heterofermentative lactobacilli. Slits are often created by heterofermentative lactobacilli when fermenting citrate, lactose, galactose, amino acids, or a mix of these substrates during the cheese-ripening process. *Lactobacillus curvatus* appears to be prevalent in cheddar cheese with the slit fault. *Lactobacillus brevis* may produce too much CO2 in cheese, resulting in unwanted gas pockets and boxed cheese blowing (Parvez *et al.*, 2006). Orange-pigmented strains of *Lactobacillus brevis subsp.* rudeness may grow in hard and white-brined cheeses. When *Lactobacillus*' fermented and *Lactobacillus brevis* concentrations are higher than usual, Italian speakers may experience blowing. Which has been shown to lack phosphotransferase systems specialized for glucose, fructose, and lactose which produce lactic acid, ethanol, acetic acid, and CO2 as byproducts of hexose fermentation via the phosphoketolase pathway. Recent research has shown that fructose-induced anaerobic growth of *Lactobacillus brevis* results in the creation of a phosphotransferase system and glycolytic enzymes that allow fructose to be processed via the Embden-Meyerhof pathway. *Lactobacillus brevis* belongs to the *Lactobacillus casei-Pediococcus* group, the lactobacilli's second phylogenetic group. *Lactobacillus brevis* has been identified from various sources, including dairy products, cheese, plants, sewage, cereals, silage, fermented meats and vegetables, cow dung, human and rat feces, and the mouth and digestive system.

Arabinose, fructose, glucose, gluconate, maltose, melibiose, and ribose are all fermented by *Lactobacillus brevis* (90 % or greater strains). Esculin, galactose, lactose, raffinose, sucrose, and xylose are fermented by 11-89 % of the bacteria. *Lactobacillus brevis* cannot grow in a chemically defined medium with pentoses as the only source of fermentable sugar. It has been demonstrated that employing carbohydrate fermentation reactions or other basic phenotypic techniques to differentiate *Lactobacillus brevis* strains is inadequate. Based on nucleic acid and biochemical evidence, *Lactobacillus brevis* strains previously assigned to the species were transferred to the new species (Novik & Savich, 2020). *Lactobacillus brevis* may be distinguished from *Lactobacillus buchneri*, *Lactobacillus hilgardii*, *Lactobacillus collinoides*, and *Lactobacillus kefir* using electrophoretic mobility of lactic acid dehydrogenases.

Calcium pantothenate, vitamin B6, thiamin, and folic acid, on the other hand, are necessary for a child's growth and development.

As a result, *Lactobacillus brevis* proteolytic activity is assumed to be reasonably modest.

It was revealed that *Lactobacillus brevis* possesses a catalase activity that is dependent on hematin as well as a nitrite reductase activity that is independent of hem.

Over 80% of cells have rounded ends and are short and straight (range: 0.7-1.0 2.0-4.0 m). Long rods, on the other hand, are more often used. They are often observed alone or in small groups. Using the Gram reaction of methylene blue stain, bipolar or other internal granulations can be detected as cells age. S-layer proteins with molecular masses ranging from 38 to 55 kDa are detected in most of immunologically different *Lactobacillus brevis* strains.

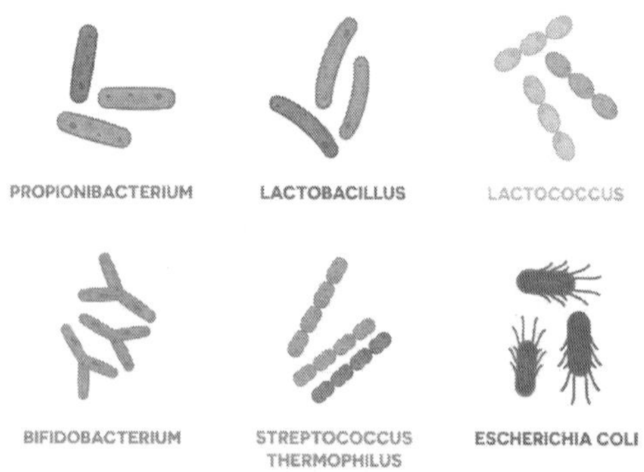

Figure 1.6: Probiotic bacteria.

Surfaces of colonies can range from smooth to rough to virtually transparent. Although some strains are orange to red in hue, they usually are non-pigmented.

1.15 LEUCONOSTOC MESENTEROIDES

When salt and low temperatures are present, a lactic acid bacterium called *Leuconostoc mesenteroides* is associated with fermentation (such as lactic acid synthesis in fermented sausage). It was connected to pathogenicity in many vegetable and food storage settings (soft rot, slime, and unpleasant odor). *Leuconostoc mesenteroides* develop tiny, grey colonies that are less than 1.0 mm in diameter and range in size from 0.5 to 0.7 m in diameter and 0.7 to 1.2 m in length. It is non-motile, non-sporogenous, Gram-positive, facultatively anaerobic, and anaerobic (Charalampopoulos & Rastall, 2009). Because its shape fluctuates depending on the media it is grown, it generally produces lenticular coccoid cells in pairs and chains. However, it occasionally generates short rods with rounded ends in lengthy chains. *Leuconostoc mesenteroides* can live from 10 °C to 30 °C, although it grows best at 30°C. It can grow at pH values ranging from 4.5 to 7.0, although it blooms at 5.5. *Leuconostoc. mesenteroides* is a form of obligate heterotactic fermentative lactic acid bacteria (LAB) that predominantly functions in industrial dairy fermentation, creating dextran, gas, and taste compounds, among other things. Its dimensions are generally 0.5-0.7 mm by 0.7-1.2 mm, and its small, grey colonies are frequently less than 1.0 mm in diameter. Lactic acid fermentation by *Leuconostoc mesenteroides* is feasible in the presence of microaerophilia. In light of this, it is vital to remember that *Leuconostoc mesenteroides*, in addition to glucose, utilize carbohydrates like sucrose and fructose as their primary energy source. As a byproduct of fermentation, it creates CO2, lactate, and ethanol.

1.16 ENVIRONMENTS

MRS agar, tomato juice agar, MRS broth, and skim milk may all be used to grow *Leuconostoc mesenteroides*. It is typically seen on the skin of many succulent fruits and vegetables. As traditional media are not appropriate for growth, tailored media are necessary to develop to a high cell density. This bacteria is widely utilized to ferment meals like kimchi, sauerkraut, and pickles with sour veggies like cucumbers and cabbage. They are included in dairy starting cultures because the bacteria can manufacture the metabolites essential for dairy product production. Citric acid CO2 and diacetyl are two of these metabolites. Diacetyl is vital to the dairy business since it is the principal source of taste and aroma many dairy products, such as buttermilk, butter, and cheeses. The formation of the eyes in certain cheeses, including Havarti, depends on CO2 production (Nichols, 2007). Customized media are frequently accepted for industrial development because traditional media, such as MRS, include meat extract and peptone that is neither kosher nor halal, which are necessary for many dairy and vegetable products. *Leuconostoc mesenteroides* grow from 10°C to 30°C, with 30°C being optimal. It may also exist at pH levels ranging from 4.5 to 7.0, with a preferred range of 5.5 to 6.5. Under aerobic circumstances, *Leuconostoc mesenteroides* have a doubling time of 0.6 h.

1.17 GENETICS

The *Leuconostoc mesenteroides* genome was successfully mapped, revealing 1762 protein-coding genes, an average genome size of 1.90138 Mbp, and a G+C content of 37.8 %.

Leuconostoc mesenteroides is a member of the Bacillota phylum and belongs to the lactic bacteria family. This is crucial because it may create lactic acid, which lowers the pH of the local surroundings and hinders the growth of competing food spoilage germs, which cannot survive in an acidic environment.

1.17.1 Health Hazards

Leuconostoc mesenteroides subsp. mesenteroides are pathogenic in a few sporadic incidences. The first human infection occurred in 1985. Penicillin and gentamicin were found to be effective antimicrobial therapy in 48 instances of a more recent hospital pandemic in 2004. The germs were identified in the patient's blood and urine (Narang *et al.*, 2011).

1.17.2 Taxonomy

There are multiple subspecies of *Leuconostoc mesenteroides.*

L. m. cremoris subsp.Garvie

L. m. subsp. dextranicum, Garvie

Mesenteroides (subsp. L. m.) Garvie

Suionicum subsp. L. m. Gu and associates

Subspecies of mesenteroides

This strain's ideal development temperature range is between 20 and 30 °C, and select strains may withstand up to 6.5 % NaCl. It has also been discovered that when yeast extract and glucose are added to milk, enough acid is created to cause the milk to curdle. Some but not all strains can ferment citrate. The demand for riboflavin, pyridoxal, and folic acid varied across the studied strains of Subsp. *mesenteroides*; need to be examined to define a starting point for the medium (Nami *et al.*, 2015). It is not needed for growth to incorporate tween 80, uracil, or combinations of uracil, adenine, and xanthine. Valine and glutamic acid are also necessary for Subsp. *mesenteroides*.

1.18 SPECIES OF CREMORIS

This strain develops between 18 and 25 °C. This strain from citrate may generate acetoin and diacetyl. Most of these subspecies' strains are unable to ferment sugar. This subspecies ferments the fewest forms of carbohydrates out of the three (McKenzie *et al.*, 2018). While galactose and maltose are strain-specific, glucose and lactose can be fermented by all strains. Riboflavin, pyridoxal, folic acid, uracil, and a mixture of uracil, adenine, and xanthine are the extra nutrients this strain needs.

1.18.1 Subspecies of Dextranicum

The ideal temperature range for this strain is between 20 and 30 °C. Glucose, fructose, lactose, maltose, sucrose, and trehalose are additional sugars that it may ferment. Also, riboflavin, pyridoxal, and folic acid must fluctuate according to the strain. Also, many strains required a combination of guanine, adenine, uracil, and xanthine. You might not be familiar with *Leuconostoc mesenteroides*, but now is the ideal occasion to bring them to the probiotics company (Mahasneh & Abbas, 2010).

In the phylum of Firmicutes, *Leuconostoc mesenteroides* is a Gram-positive bacterium. It is neither spore-forming nor non-motile, meaning it cannot move independently. The *Lactobacillales* order, to which it belongs, is the same as that of the *Lactobacillus*, but after that, the two bacteria are different. It belongs to the family *Leuconostocaceae*.

All other strains are compared *to Leuconostoc mesenteroides subsp. mesenteroides* type strain.

Cremoris, dextranicum, mesenteroides, and suionicum are the four separate subspecies. It is a facultative anaerobe as a species.

These bacteria usually form chains or pairs and have spherical forms. They are a vital component of the fermentation that gives kefir, certain cheeses, and sauerkraut their particular tastes.

1.18.2 How harmful are Leuconostoc mesenteroides?

Although these bacteria are infrequently able to infect immune-compromised people, they are typically considered beneficial microbes. In one case study involving deaths, it was found that two independent outbreaks over the years were brought on by tainted parenteral (food-tube feeding) or liquid sustenance, both of which were prepared in the same hospital pharmacy in Spain. Although the folks consuming the food were pretty unwell happily, not all of them finally perished from the illness (Li *et al.*, 2016).

Leuconostoc mesenteroides: Is it a Probiotic?

Some strains of this bacterial species are shifting from helpful bacteria to probiotics as more inquiry is made about them. *Leuconostoc mesenteroides subsp. mesenteroides* NTM048 is one such strain that was discovered in green peas. It has been established that NTM048 operates as a probiotic and modulates immunological responses throughout the body by boosting sIgA. The immunological protein known as sIgA guards mucosal surfaces, particularly the digestive system, against intruders such as harmful bacteria and viruses (Lin et al., 2019). NTM048 can enhance sIgA levels and the immune system's reaction, protecting from infections and maintaining intestinal cells healthy and next to one another.

Another strain creates a prebiotic, which is not considered a probiotic in and of itself but has been identified to let certain Bifidobacterium and Lactobacillus proliferate. *In vitro* investigations of this prebiotic discovered that it had an inhibitory action on human colon cancer cells.

1.18.3 Where are Leuconostoc mesenteroides found?

Many different kinds of fresh vegetables have *Leuconostoc mesenteroides* on them. Pathogens in ferments and the body are decreased or destroyed by the acids and bacteriocins they generate.

In addition to the meals mentioned above and beverages, Raw Probiotics for Women is at least one probiotic supplement that contains *Leuconostoc mesenteroides*.

To assist and obtain the probiotics the gut craves, foods like live-cultured yogurt, sauerkraut, miso, kimchi, and kefir maybe added to the diet or create a wonderful probiotic smoothie. The diet can also be complemented with a high-quality probiotic to strengthen your immune system, digestive system, and general health. Complete Probiotics Platinum® by 1MD Nutrition contains 11 unique strains and offers 51 billion live cultures in each capsule for the best possible support of the digestive and immune systems.

Check nutrition labels to verify purchasing a healthy product, as many foods include high quantities of added sugar, salt, and fat. When fermenting at home, make sure to adhere to the recipes strictly. Food may rot and become unsafe to eat as a result of improper temperatures, extended fermentation periods, or sloppy equipment (Kumar, 2013). Flatulence and bloating may be the first adverse side effects of fermented meals. Always check the nutrition labels on shop items and follow the instructions when fermenting at home to avoid spoiling.

CHAPTER 2

PROBIOTIC CHARACTERISTICS OF BACTERIAL STRAINS AND YEAST

CONTENTS

2.1 INTRODUCTION

One of the most diverse microorganisms on the planet, yeasts have a wide range of species. Non-food applications are possible because of their wide range of biological activity. While they play a critical part in the flavor creation of fermented foods, it has long been documented that they have a detrimental effect on pathogenic bacteria and fungi. Antifungal killer toxins (also known as "mycocins") and antibacterial chemicals, as well as the acidity of their growth media and their ability to tolerate high ethanol concentrations, are all associated with these traits (Kukkupuni *et al.*, 2015). While *Lactobacillus* and *Bifidobacterium* have dominated the development of probiotic foods, *Saccharomyces cerevisiae var. boulardii* has long been known to be beneficial in treating gastroenteritis and other gastrointestinal disorders.

The antibacterial properties of yeasts are examined in this article. We will talk about the processes at work here, as well as the current medical and veterinary applications for these physiologically active yeasts. The name "yeast" came from the Dutch word "gist," which refers to froth that forms during the fermentation of beer wort. The French word "levure" refers to the action of yeast in allowing bread dough to rise (Kim & Park, 2017). Yeasts are one of the the most critical microorganisms in history, commerce, and science. Beyond fermented foods and beverages, yeasts have several other uses, including animal feed and veterinary operations, medicine, the biomedical industry, and pharmaceuticals. Yeasts, moulds, and bacteria are among the primary targets when fighting other microorganisms.

The original discovery of yeast inhibiting activity came from Secondary metabolites known as "killing poisons" or "mycocins" which were later shown to be responsible for the yeasts' aggressive behaviour toward other yeasts. In evaluating the antimicrobial activity of yeasts in fermented foods and drinks, Viljoen (2006) highlights the impact of organic acids, antibiotic factors, volatile acids, hydrogen peroxide, and a range of other substrates discharged from the product. However, relatively little work has been performed on the mechanisms of yeast inhibition.

Although only *Saccharomyces cerevisiae var. boulardii* has been studied in depth and its inhibitory mechanisms have been well-defined, significant antagonistic activities have been associated with several other genera or species, and it is anticipated that their potential industrial and biotechnological applications will be investigated.

2.2 YEASTS AND BACTERIAL STRAINS UNFRIENDLY TENDENCIES

The employment of antagonistic bacteria to inhibit pathogenic bacteria has been the object of considerable research over the years, but yeasts have attracted less attention for a similar reason. Consequently, the research and future application of antibacterial compounds created by yeasts are in their infancy. Antagonism of microorganisms by yeasts has been primarily attributed to competition for nutrients, pH changes in the medium due to growth-coupled ion exchange or organic acid production, production of high concentrations of ethanol, and secretion of antibacterial compounds and release of antimicrobial compounds such as killer toxins or "mycocins." Mycocins are extracellular proteins or glycoproteins that impact the cell membrane's function in yeasts with receptors for the chemical. Their activity is primarily aimed at yeasts closely linked to the producer strain, which has a protective component (Khoruts, 2018).

In the brewing business, the first mycocins were identified in association with S. cerevisiae. Several have since been isolated, often in high-density yeast colonies under intensely competitive circumstances. Numerous yeast genera, including Saccharomyces, Candida, Cryptococcus, Debaryomyces, Kluyveromyces, Pichia, Torulopsis, Williopsis, and Zygosaccharomyces, produce mycocin. The deadly toxin phenotype may be carried on extrachromosomal elements in the form of double-stranded RNA viruses on double-stranded linear DNA or a chromosome. Well-known mechanisms of the killer toxin include the interruption of cell division by inhibiting DNA synthesis, the inhibition of synthesis of the cell wall component -1, 3-glucan, and ion leakage resulting from the formation of channels on the cytoplasmic membrane. Unlike yeast-against-yeast antagonism, yeast antibacterial capabilities have received much less attention.

In theory, probiotic cells should have the features of a large population capable of surviving the gastrointestinal system and colonizing the intestinal epithelial cell lining for an extended length of time to give health advantages to the host. This study assessed the probiotic properties of four isolated bacterial strains and yeast based on their varied probiotic traits (Jugan *et al.*, 2017). Previously, *Leuconostoc plantarum* FH185 has exhibited good survival under low pH and bile salt conditions, 97.4 %, and 96.66 %, respectively. Bacillus sp. T3, T4, SM2, and JSP1 demonstrated decreased cell viability at pH 2 below 65 % but above 80 % bile salt resistance. In the current experiment, 97 % of all strains demonstrated significant pH 2.5

resistance (P0.05). However, tolerance to bile salts varies throughout strains, with S. cerevisiae CP605 cells displaying the highest tolerance. Although Lactobacillus sp. CP134 and Bacillus subtilis CP350 cells displayed a high tolerance for bile salts, *Lactobacillus sp.* CP133 cells exhibited no viability, suggesting that bile salt negatively affected cell proliferation in this strain. Bile salts induce structural instability of the cell membrane, resulting in cell content leakage and death, as proved by the entire demise of CP133 cells in 0.3 % bile salt. The thermotolerant behavior of the strains depicted in Figure 2 demonstrated CP350 and CP605's sustained resistance to temperature increases. In contrast, the tolerance rates of *Lactobacillus sp.* CP133 and CP134 fell as the temperature ascended. In general, the protective matrix in the microbial cell wall is destroyed at high temperatures, indicating that *Lactobacillus sp.* CP133 and CP134 have poor tolerance rates (Johnston *et al.*, 2012). Moreover, *Bacillus subtilis* was reported to generate heat shock proteins in response to an increase in temperature, validating the thermotolerance of CP350. All of the isolates tested in this study demonstrated thermos-tolerance at 30°C and 37°C, regarded as the mammalian gut environment and rapid growth time for industrial production.

S. cerevisiae CP605, Lactobacillus sp. CP133 and *Bacillus subtilis* CP350 adhered differently to the HT-29 monolayer than the reference strain *Lactobacillus rhamnosus* GG. In addition, the S. cerevisiae strain studied in this work (CP605) demonstrated comparable adhesion values to the well-known probiotic bacterium *Lactobacillus rhamnosus* GG, indicating that it may be an effective *in vivo* colonizer. Most of the time, adhesion to mucosal cells is crucial for synthesizing enzymes, lactic acid, vitamins, and antibacterial compounds (Hill, 2010). Intriguingly, all strains in the current experiment demonstrated greater adhesion to HT-29 monolayer cells than reference strains. Nonetheless, pH, incubation length, and cell concentration are crucial for the ability of probiotic cells to bind to the gut mucosa.

The antibacterial activity of Lactobacillus sp. CP133 and CP134 in the current experiment demonstrated a broad capacity to inhibit the growth of bacterial pathogens, which was consistent with previous findings; regarding the antibacterial activity of Lactobacillus sp. isolated from fermentation conditions. As previously observed, the inhibitory power of Lactobacillus sp. CP133 and CP134 can be connected with the synthesis of bacteriocin-like metabolites. Lactic acid bacteria are known to hinder the growth of pathogenic organisms. In our analysis, neither *B. subtilis subsp. subtilis* CP350 nor *S. cerevisiae* CP605 demonstrated antibacterial activity against the examined hazardous microorganisms.

Lactobacillus sp. exhibits resistance to glycopeptides, aminoglycosides and quinolone antibiotics. *Lactobacillus paracasei* CP133, and *Lactobacillus plantarum* CP134 were shown to be resistant to kanamycin, metronidazole, and vancomycin in the present examination, which is consistent with other findings. Additionally, Lactobacillus sp. CP133 revealed resistance to trimethoprim and transitory tolerance to gentamycin, although *B. subtilis subsp. subtilis* CP350 exhibited resistance to amoxicillin and clindamycin. The remaining tested antibiotics decreased the growth of the bacterial strains. Cell development inhibition is commonly connected with reduced protein production in target strain cells.

The robustness of the cell wall of probiotic strains to the antimicrobial hydrolytic enzymes found in lysosomes is critical for preventing breakdown. *Bacillus subtilis subsp. subtilis* CP350 demonstrated high resistance to lysosomal activity among the bacteria. It has been established that resistance to cell wall degradation by lysosomal hydrolytic enzymes in *Bacillus spp.* is associated with these organisms' intrinsic potential to create endospores (Harzallah & Belhadj, 2013). In addition, cell wall resistance to breakdown by lysosomal enzymes occurs if the cell wall's functional groups persist in the N-acetylation state, whereas O-acetylation promotes destruction. Proteases from microorganisms are vital for protein metabolism, cell survival, and identification. As expected, *B. subtilis subsp. subtilis* CP350 demonstrated high proteolytic activity. Proteolysis has detected Bacillus bacteria because they are the principal producers of proteases, especially alkaline proteases.

In this study, microbial strains CP605, CP350, CP133, and CP134 were evaluated for their potential to diminish odor in addition to identifying probiotics. The odor caused by the decomposition of pig manure is the cause of complaints from neighbors and an issue of public concern. Recently, bio-additives and microbial products have been utilized to decrease cow odor. During decomposition, nitrogenous and short-chain fatty acids are formed due to a lack of oxygen and a high temperature. Using concentrations after 24 hours of treatment, the odor-reduction efficacy of the microbial strains was examined. The results for all bacterial strains varied from 10.2 % to 45.5 %, which was lower than the decrease rates recorded for reference strains. These findings showed that these microbial strains may be valuable in the sector of odor control.

Existing commercial strains of probiotic bacteria are mainly sourced from the terrestrial environment. Marine yeasts are phenotypically unique from terrestrial yeasts and are highly resistant to growth inhibitors such as salt, acid, and temperature. Marine microorganisms, particularly marine yeasts, have exhibited potential probiotic characteristics. *S. cerevisiae*, a marine yeast, was highly resistant to salt, with higher glucose uptake for greater fermentation capacity (Gwee *et al.*, 2018). As the use of probiotics in aquaculture grows increasingly common, researchers have explored the probiotic characteristics of bacteria isolated from the ocean. *Lactobacillus plantarum* recovered from fish gut boosted tilapia immune system, resulting in disease resistance, and bacteriocinogenic bacteria isolated from marine species were examined for probiotics properties for application in the feed industry. *S. cerevisiae* CP605 isolated from farm soil and *B. subtilis subsp. subtilis* CP350 isolated from cow slurry displayed substantial resistance to temperature, acidity, and bile salt stresses, which is a favorable characteristic for probiotics. Kimchi-isolated *Lactobacillus plantarum* CP134 has proven the capacity to adapt to extreme stomach conditions. Future studies will incorporate animal experiments to evaluate the influence of all isolates on animal health and productivity. In addition, all isolates demonstrated a considerable decrease in the odor components of swine excrement. The data show that strain-specific functional probiotic characteristics and odor reduction capabilities exist. Combining of these microbial strains can potentially provide valuable resources for probiotics and odor reduction businesses.

2.3 ACID AND BILE TOLERANCE AND INTESTINAL ADHESION ACTIVITY

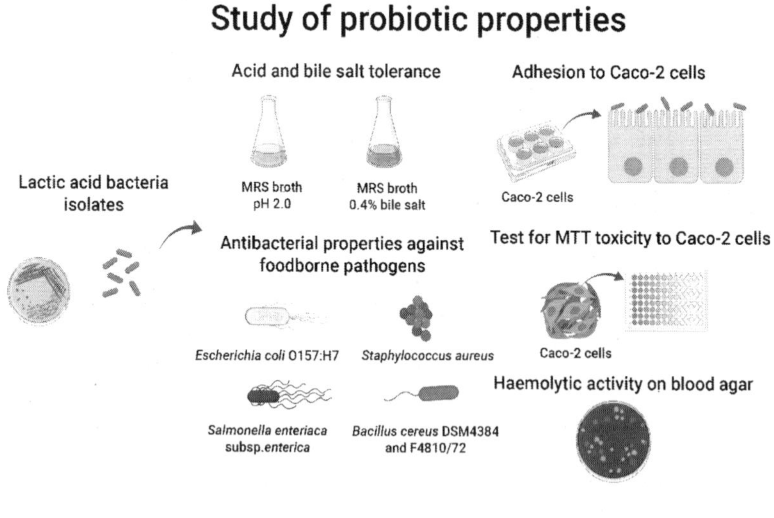

Figure 2.1: Some properties of probiotics.

Probiotic bacteria must be acid and bile-resistant, two of the essential qualities. A pH of at least 3.0 is commonly accepted as the minimal need for a culture to be considered effective as a probiotic. This is the most often used way of testing for determining probiotic acid tolerance. This experiment, used probiotic supplementation to study gut tolerance to bile salt and low pH. After two hours of incubation at a pH of 2.5, all bacteria exhibited a viability rate greater than 97 %. After an 8-hour incubation period with CP134 at a concentration of 0.3 %, a more than 95 %tolerance rate was found (P0.05). However, CP133 demonstrated exceptional acid tolerance after being treated with bile salt for 8 hours, no bacteria were present (Cammarota et al., 2014). CP605, on the other hand, demonstrated tolerance to acid and bile of more than 97 %. The viability of *S. cerevisiae* ATCC18824, *B. subtilis* ATCC6051, and *Lactobacillus rhamnosus* GG ATCC53103 was examined in conditions of low pH and 0.3 % bile salts as a positive control.

Lactic acid bacteria supplements have been demonstrated to be beneficial, and food additives are a popular source of these bacteria. According to the literature, some of the data regarding their reported advantages is conflicting, and it is unclear how they operate. Variability is increased due to the frequent use of many strains in research lacking sufficient quality control techniques. Many organisms do not have information on their acid shock or acid-tolerance response, which are known to differ throughout bacterial species. The development of acid-resistant and bile-resistant strains of *Lactobacillus acidophilus* and an analysis of the phenotypic alterations were our key goals.

By exposing the bacteria to hydrochloric acid (with a pH range of 3.5 to 7.0) and mixed bile salts in fast succession, acid- and bile-resistant isolates were generated by natural selection processes. The acid- and bile-tolerant isolates kept their capacity growing at a pH of 3.5 with 0.3 % bile even when the selection pressure was eliminated and restored. Isolates are distinct from their parents in many aspects, including their resistance to freezing and lactose, their levels of protease and aminopeptidase activity, the profiles of their plasmids, and the fatty acid profiles of their cell walls. These findings reveal that the created acid- and bile-tolerant isolates have improved growth properties than their parent strains when challenged with a challenging environment. Furthermore, these isolates have the potential to be regarded as candidates for probiotic strains following further characterization employing animal models (Gogineni *et al.*, 2013). The bacteria categorized as lactic acid bacteria are commonly referred to as probiotics, and they are added to meals as dietary supplements to deliver a wide range of health advantages.

These bacteria, primarily *Lactobacilli* and *Bifidobacteria* may include several medical advantages, such as antibacterial action, anticholesterol activity, increased lactose utilization, and anticarcinogenic activity. There is minimal evidence to show that animal models exposed to probiotic micro-organisms experience improvements in their overall health. Renner claimed that the paucity of convincing data on the biological activity is attributed to the insufficient experimental design and the variation in strain settings employed in clinical research. Even if this is true, it is evident that additional study is essential to firmly verify the degree of the health advantages that diverse strains offer to animals. Bacteria exploited as probiotic adjuncts are commonly given via a food system. The germs begin their voyage to the lower intestine system by entering the body through the mouth because of this.

As a result, the bacteria that make up probiotic supplements should be resistant to the enzymes present in the mouth (such as lysozyme). They should also be able to survive the digestive process in the stomach and intestinal systems. According to Berrada et al., it takes for a medication to transit through the stomach and be emptied 90 minutes (Gupta, 2011). However, other digestive processes have a longer residency duration; thus, the bacteria need to be resistant to the harsh environment of the stomach and upper intestine, which contain bile. This is because more digestive processes have longer residence durations. The pH in the stomach can drop as low as 1.5, which is when the first signals of cellular stress manifest.

After the germs have gone through the stomach, they reach the upper intestine tract. It is here that the bile is released into the intestines. It is not viable to generate a precise estimate of the quantity of bile that is present in the human digestive tract as it is constantly changing. After surviving in such a challenging environment, the bacteria can colonize the epithelium that lines the lower intestine tract. Therefore, before probiotic bacteria may begin to deliver any health advantages, they need to be able to endure acid for at least 90 minutes, tolerate bile acids, attach to the epithelium, and grow in the lower digestive tract. Only then may probiotic bacteria begin to deliver any health advantages. This investigation identified isolates of *Lactobacillus acidophilus* resistant to their environment (Gao et al., 2022). In determining the phenotypes of the isolates, significant focus was devoted to finding features critical to the physiology of lactic acid bacteria.

2.4 CHOOSING ISOLATES CAPABLE OF WITHSTANDING ACIDITY

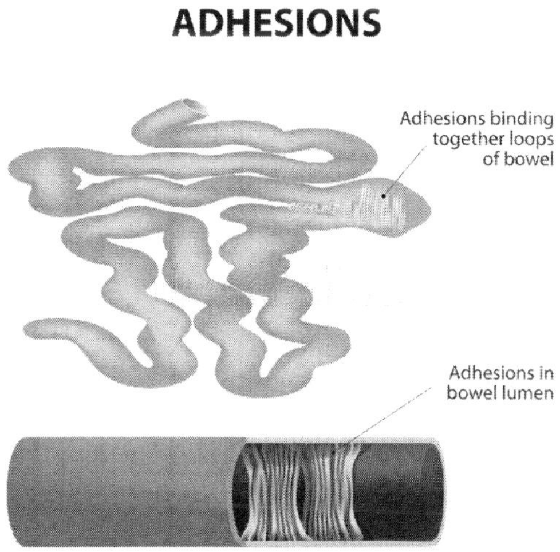

Figure 2.2: The structure of the intestine and its lumen.

Before being utilized in an experiment, each parent strain of *Lactobacillus acidophilus* was cultured once in MRS broth received from the stock freezer vial for a duration ranging from 16 to 24 hours at a temperature of 37 °C. Following this incubation, the cells were collected by centrifugation (4300 g for 10 minutes at 4 °C), washed three times in sterile saline, and then inoculated into MRS broth acidified with concentrated hydrochloric acid to pH 3.5 or MRS broth that had not been acidified. Afterward, the cells were incubated in a temperature-controlled spectrophotometer at 37 °C (Faujdar et al., 2016). Using the pour plate technique, plate counts were collected before and after the incubation period using MRS agar with a pH of 6.8. The selection criteria for acid-resistant strains were based on strains that displayed little to no drop in the number of colony-forming units produced per milliliter of culture. After that, these bacteria were utilized to separate ten unique colonies from the acidified MRS agar.

Each acid-tolerant candidate was sent to additional testing to determine whether or not it has the capacity to develop in an acidic environment. This was achieved by streaking the bacteria over acidified MRS agar. The plates' growth was studied after 24 to 96 hours of anaerobic incubation at 37 °C. At

37 °C, selected colonies were cultured in an MRS broth with an acidic pH of 3.5 for 24 to 96 hours. The colonies that were confirmed to be resistant to acid and bile were utilized in the completed isolation experiments so that extra investigation could be conducted.

2.5 SELECTING ISOLATES CAPABLE OF WITHSTANDING BILE

sMRS agar was utilized in the direct plate test devised by Christiaens et al. to examine each acid-tolerant isolate's level of bile resistance. At pH 4.0, 5.0, 6.0, or 7.0, 0.3 % of the pH-adjusted agar contained glycocholic acid, glycodeoxycholic acid, taurocholic acid, taurodeoxycholic acid, or oxgall. All of these bile acids were revealed to include oxgall as well. The precipitation of bile salts prohibited further testing below pH 4.0 to deconjugate bile salts. After streaking each acid-resistant isolate onto plates containing different bile salts on MRS agar at each pH level, the plates were put in an anaerobic environment and cooked to 37 °C for 24 to 96 hours. The growth and bile precipitation on the plates were evaluated at intervals of 24 hours each (Ezema, 2013). Where there was growth, ten separate colonies of each strain were selected and injected into MRS broth containing 0.3 % of the prescribed bile salt at the exact pH from which the isolate was taken. If the isolate was proved to be able to grow in this broth, then it was determined to be acid- and bile-resistant.

2.6 AN INCREASING NUMBER OF ISOLATES CAPABLE OF WITHSTANDING BILE AND ACID

After being frozen at -70 °C for 48 hours, the acid- and bile-resistant isolates were allowed to defrost at room temperature (25 °C for around five minutes) before being tested for their potential to proliferate. Thawed isolates were cultured for 16 to 24 hours in their respective isolation broths at 37 °C to investigate the extent to which the relative cell density altered over the length of the incubation procedure. The difference in the A650 measurements collected before and after the incubation time was utilized as a metric to determine the degree to which the cell density had moved. The strains that developed at the fastest speed were selected for the ensuing investigation.

Additional attempts at cell growth were made to simulate passage through the intestines. The cells were cultured in their respective medium before being collected by centrifugation (4300 g for 10 minutes at 4 °C), washed three times with PBS (pH 7.0), and then resuspended in the same quantity of saline (pH 7.0) as the initial culture. The cells were then let return to their original culture. It was selected to infuse one % of the initial component of the suspension into both the acidified (pH 4.5) and the unacidified (pH 4.5) MRS broths (pH 6.8). A650 readings were acquired at regular intervals of 15 minutes to quantify the growth rate of the cultures cultured in a spectrophotometer with temperature control at 37 °C for an hour and a half (Deng et al., 2020). The pH 4.5 acidified MRS broth was injected with 1 % of the remaining suspension, and the A650 was measured every 15 minutes during the operation. After 90 minutes of incubation at 37 °C, each bile salt was added to the broth until the final total concentration of 0.2 % was obtained, with individual bile salts having the same concentration. After that, the pH of the broth was increased up to 6.8 using sterile 1N sodium hydroxide. After making different modifications to the pH and bile salts, the A650 was investigated at 37 °C for an additional 510 minutes at 15-minute intervals.

2.7 SELECTING ISOLATES CAPABLE OF WITHSTANDING BOTH BILE AND ACID

To isolate bile-tolerant variants, each parent strain (L. acidophilus ATCC 43121, 33200, and 4962) and acid-tolerant isolates from these strains were inoculated onto MRS agar with a variety of bile salts at a range of pH values (pH 4.0 to pH 7.0). (pH 4.0 to pH 7.0). (pH 4.0 to pH 7.0). Both growth that was bile salt deconjugation dependent, as indicated by a white precipitate surrounding the colony, and growth that did not need bile salt deconjugation was identified. Regardless of the criterion applied for selection, there was no evident pattern demonstrated to be predictive. After an incubation period of 24 hours, most bacteria flourished at pH values ranging from 5.0 to 7.0 in the presence of each bile salt (Deshpande et al., 2011). This experiment aimed to examine how effectively growth would occur at a pH of 4.0 when bile salts were present.

Although glycocholic acid and oxgall successfully inhibited all of the isolates tested in conditions with a pH of 4.0, glycodeoxycholic acid successfully promoted the growth of most organisms. As a direct result, colonies of each acid-resistant isolate were chosen To differentiate them from one another, they were each given the label GD4, followed by their unique isolate numbers. For example, the colonies from the acid-resistant isolate LSC2 (derivatives of the parent ATCC 43121) developed in MRS agar with glycodeoxycholic acid at pH 4 were given the names LSC2-1 GD4 and LSC2-2 GD4, respectively. These designations were allocated to the colonies in order of their development in the medium (Table 3). As a result of these results, we came up with the working hypothesis that it may be able to generate colonies resistant to bile and acid. These experiments also indicated that the phenotypic features of probiotic bacteria might change throughout the digestion process, resulting in a unique population in the digestive system from that which was initially ingested.

Therefore, exhaustive strain characterization is required to determine the phenotypic characteristics of acid- and bile-tolerant isolates. In contrast to Lankaputhra and Shah's conclusions, our data demonstrated that bile-resistant isolates developed over time when they were treated with lengthy incubation periods. Acid and bile levels were evaluated throughout the analysis's screening process, indicating a substantial departure from earlier studies in this field. Even though the experiments were designed in various ways, they all arrived at the same finding: bile and acid both had an individual and combined influence on the multiplication of bacteria (Di Cerbo & Palmieri, 2015). Based on these findings, it is recommended that an acid- and bile-resistant probiotic bacteria supplement be given to dairy products to help prevent phenotypic abnormalities that are not specified after ingestion.

2.8: THERMO-TOLERANCE

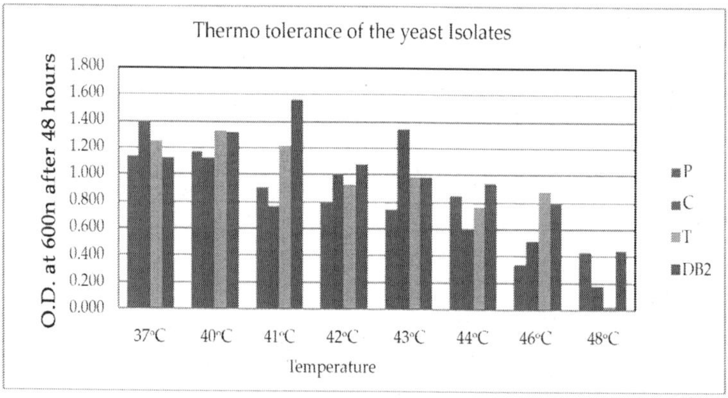

Figure 2.3: Thermo tolerance.

Comparatively, CP605 and CP350 demonstrate excellent tolerance to temperature rises compared to reference strains, as evidenced by the strains' high tolerance to temperature rises. When it comes to temperature, however, the Lactobacillus strains that we studied were robust to temperatures as high as 60°C. Due to solid tolerance to industrial production settings; no strains were resistant in both, no strains were found to be resistant to both conditions (Daliri & Lee, 2015).

Several processes and variables contribute to heat resistance or thermotolerance, including stress proteins, molecular chaperoning, chaperonins, chemoprotective chemicals, innate protein thermostability, cell wall composition and architecture, multicellular structure creation, and spore production.

A consequence of heat stress is protein denaturation. The cellular proteins denature and agglomerate when the degree of stress is too high (a heat-shocking temperature significantly above the OTG or a very prolonged heat shock). The result is cell death. A hard-boiled egg is a good example. Denatured, aggregated proteins cannot be separated from one other and re-assume their natural, functional structure while trying to revive hard-boiled eggs. On the other hand, protein denaturation seldom reaches such a severe degree that it cannot be repaired. Weaker stressors partially denature the protein, but the damage is repairable, so the protein can return to its original functional state. Molecular chaperones are one of the many mechanisms that play a role in recovery. In addition, these chaperones limit protein breakdown and remove denatured protein molecules.

One of the essential mechanisms supporting thermotolerance is the ability to avoid and cure protein denaturation, and eliminate irreparably damaged and aggregated proteins before they interfere with intracellular machinery.

The term "thermotolerance" refers to a cell's or organism's ability to withstand potentially deadly heat stress. The phrase "heat hardening" has garnered the most significant attention. However, this capacity can evolve through several processes, including changes in growth status (Elliott et al., 1996) and long-term acclimatization to higher temperatures (Laszlo and Li, 1985; Cavicchi et al., 1995). Tolerance to high, but not fatal, temperatures is one type of thermotolerance. After a time of recovery, conditioned cells (or organisms) display a transitory resistance to a heat shock that kills most unconditioned cells. In cultured cells and organisms, changes in HSP levels have been linked to the establishing and losing thermotolerance.

Temperature tolerance may be induced in cell lines by overexpressing HSP27. Thermotolerance may be developed in a variety of ways. Heat shock damages many cellular structures and processes, and even in cultured cells, establishing thermotolerance by means other than heat shock or over-expression of a specific HSP protects many of these structures and functions (e.g., protein synthesis, rRNA transcription and processing, mRNA splicing, microfilament, and nuclear integrity). Although these "partially protected" states are unlikely to be identical to the thermotolerant state induced by heat shock, which shields numerous targets, they may be. Any goals that may be safeguarded increase one's chances of surviving.

At the level of the organism, increased heat resistance is an even more complicated phenomenon. Acute exposure to temperatures on the verge of mortality produces critical thermal maxima, which have been used frequently in studies of species differences in thermal resistance. This technique often inhibits heat-induced reactions. It takes a long time for HSPs to build up at high temperatures due to the inhibition of protein synthesis, which delays the recovery phase.

Those responsible for this inherited intrinsic temperature tolerance are still unidentified genetic loci. Scientists have discovered differences in thermal resistance between closely related species (Bosch *et al.,* 1988; Sanders *et al*., 1991; Hoffmann and Parsons, 1991) and between populations within a species using ecologically relevant thermal stress regimens (Norris *et al.,* 1995; Krebs and Feder, 1997b). Higher HSP expression was linked to greater resistance to temperature stress in some of these studies (Bosch *et al*., 1988; Sanders *et al*., 1991). There was a positive connection between HSP70 synthesis and temperature tolerance in individuals from outbred populations of the live-bearing fish *Poeciliopsis gracilis* (Norris *et al*., 1995). Quantitatively, it has been shown that both HSC70 and HSP70-induced thermotolerance may be attributed to Poeciliopsis hybrids (diIorio *et al*., 1996). HSP70 expression and acquired heat resistance hurt *Drosophila melanogaster* survival rates without stress (Krebs and Feder, 1997b). Strains that collect higher amounts of HSP70 as heat shock may do the same during early development, when synthesis is accelerated without stress, with potentially harmful outcomes. Trade-offs between negative and positive effects may sustain the variety of HSP70 concentrations in populations. Heat shock response in organismal biology is now well summarized in Feder and Hofmann's latest review (1999).

2.8.1 Thermal Resistance Is Influenced By a Variety of Factors

The pace at which bacteria and their spores die when exposed to heat is influenced by environmental conditions. Strains of the same species can differ in heat resistance. A drop in water activity enhances thermal resistance substantially. This is especially true for high-sugar, protein, and fat meals. However, a low pH and an acidic environment diminish heat resistance substantially. When heating a product, the product's pH is of utmost importance. Products with a pH below 4.5 can be pasteurized at temperatures as low as 100 °C, but foods with a pH over 4.5 must be sterilized at temperatures over 100 °C. At pH 4.5, C. botulinum, the most prominent pathogenic endospores bacterium, cannot grow or generate toxins, and any spores that survive heat treatment are incapable of germination (Dai et al., 2018). In the canning process, the interaction of heat with different preservatives can be employed to benefit.

2.9 ANTIBACTERIAL ACTIVITY AND ANTIBIOTIC SENSITIVITY

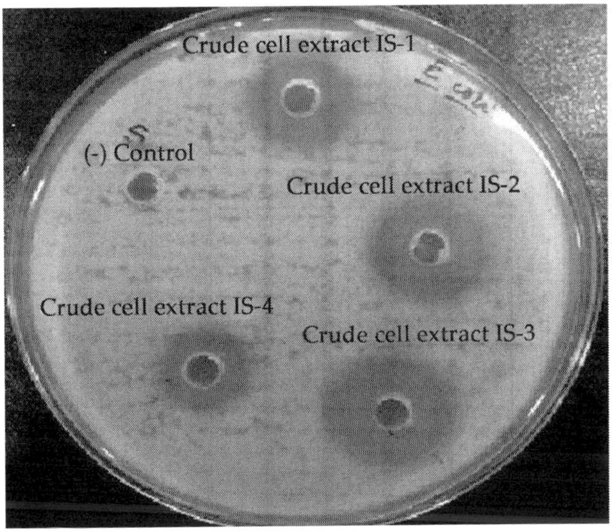

Figure 2.4: Antibacterial activity.

Astonishing results were obtained about the studied diseases using the CP133 and CP134 strains. CP134 exhibits a distinct inhibitory zone when tested against Y. pseudotuberculosis NCCP11125 and *Escherichia coli* KCTC2617. On the other hand, the indicator pathogens CP605 and CP350 had no effect. E-strips were used to test the antibiotic sensitivity of the strains, and the results are shown in Table 4 below. An antibiotic-resistant superbug was already in full bloom at this point. CP133 and CP134 were both resistant to only Trimethoprim antibiotic. Amoxicillin and clindamycin tolerance was required for CP350 to develop resistance to vancomycin, whereas the latter was only able to do so after a brief period.

Microbes can wreak havoc on any food, even in industrialized countries like the United States. 40 % of the world's food supply is lost or wasted each year due to microbial degradation, according to the United Nations (Gustavsson *et al.,* 2011). Examples of microorganisms that threaten meals and food products include yeast, mold, and bacteria, which can wreak havoc (Lianou et al., 2016). When these bacteria are in the food, they begin to multiply and produce substances that aid in the digestion of the food (Parlapani *et al.,* 2017). Foodborne illness, which occurs when someone becomes ill after consuming tainted food, is one of the most common dangers to food safety (Azziz-Baumgartner et al., 2005; Kirk *et al.,* 2017).

Microbes from the surrounding environment are utilized throughout the harvesting, slaughtering, processing, and packaging processes (Hatab *et al.,* 2016). Conventional pasteurization and low temperatures do not affect these microorganisms (Dimitrijevi et al., 2007; Provincial et al., 2013); Saraiva *et al.,* 2016; Sade *et al.,* 2017). The use of synthetic chemicals in food preservation has decreased as more people become aware of the potential dangers they pose to their health (Gyawali and Ibrahim, 2014; Caleja *et al.*, 2016; Kalem *et al.,* 2017). New technologies that are less harmful to the environment will have to be developed to safely preserve food and increase its shelf life because chemical preservatives are no longer available. Fernández-López and coworkers, Suppakul and coworkers, and Clark and coworkers studied natural preservatives derived from plant extracts in 2005.

Various parts of plants have been used for centuries for treating a wide range of ailments in humans, (Khan *et al.*, 2013). There are several chemicals in medicinal plants that are either antibacterial or antioxidant (Talib and Mahasneh, 2010). Antibacterial properties of a wide range of plant species have been studied extensively. However, Gram-positive and Gram-negative bacteria are resistant to the antibacterial properties of cinnamon, garlic, herbs such as basil and curry, and spices such as ginger, sage, and mustard (Alzoreky and Nakahara, 2003; Castro et al., 2008). As previously reported, Chinese chives and cassia extract effectively inhibit the growth of *E. coli* and other pathogens when meat, juice, and milk are preserved with these ingredients (Mau et al., 2001). An alcoholic extract of curry leaves was administered 48 hours later, and the growth inhibition of *Candida albicans* produced a value of 24.05.0.07. According to Nzeako et al., thyme oil extract effectively inhibited the growth of *Pseudomonas aeruginosa* and *Candida albicans*. (2006).

The antibacterial activity of medicinal plant extracts is the first step in improving their use as naturally occurring antibacterial agents to protect food quality and extend shelf life. Extracts of thyme, rosemary, clove, and roselle were evaluated against seven common diseases and spoilage microbes in a study published in the Journal of Agricultural and Food Chemistry. Listeria monocytogenes, *E. coli* O157:H7, Salmonella, and *Staphylococcal aureus* bacteria were present. The antibacterial activity was discovered in these extracts.

Antibiotic susceptibility testing is helpful in the study of infectious diseases, the discovery of new drugs, and the improvement of therapeutic results. Antimicrobial activity was the primary focus of our testing protocols when evaluating the antibacterial activities of pure medicines and extracts.

Most important antibiotic classes were discovered and investigated during the so-called "golden time" of antibiotic discovery in the 1960s (tetracyclines, cephalosporins, aminoglycosides, and macrolides). History repeats itself, and these intriguing chemicals could probably lose their usefulness due to increased microbial resistance. Public health is concerned about multidrug-resistant microorganisms because they can make treatment ineffective

As a result, the development of new antibiotics is essential. Natural materials continue to play an important role in synthesizing novel medicinal molecules. This bacterium may be found in eukaryotic microorganisms, plants and some prokaryotic organisms. Antibacterial compounds derived from microorganisms and plants have made up most chemicals.

It is possible to find a wide range of structural complexity in plants and natural chemical sources. Antibacterial agents can be found in secondary metabolites, essential oils, plant and microbial extracts, and newly produced chemicals (Chugh & Kamal-Eldin, 2020). Since there is no standardization in the methods employed to create the inoculum, the amount utilized, the growth medium, the incubation settings, or the endpoints in published studies on the antibacterial activity of natural substances, no inferences can be drawn from these studies.

Plant extract antimicrobial activity data must be accurate and comparable to prevent studies in which antimicrobial activity is only examined due to the phytochemical analysis.

In vitro antibacterial activity testing of an extract or a pure medication can be carried out using a variety of scientific approaches. Two methods are used almost exclusively: agar dilution and disk diffusion. Antifungal drugs are tested using various methods, including eating contaminated food. An in-depth examination of an agent's antimicrobial effect is best accomplished using the time-kill and flow cytofluorometric techniques, which show the type of inhibitory effect (bactericidal or bacteriostatic) and the degree of cell damage that the test microorganism suffers as well as whether the effect is time-dependent or concentration-dependent. An antimicrobial sensitivity or susceptibility test can be used to determine if a bacteria is resistant to an antibiotic. Because some drugs are ineffective in treating specific diseases, this strategy is employed. For clinicians, using directed therapies allows them to shift away from empiric therapies that rely on a physician's clinical suspicions of an infection's location and the most likely causative bacterium to choose an antibiotic.

Sophisticated approaches such as genetic or culture-based sensitivity testing expose microorganisms to drugs and generate resistance. Conventional practice defines the size of inhibition zones on agar plates that have been evenly loaded with bacteria for culture operations. These are regions surrounding paper discs with antibiotics that hinder bacterial growth. This may be used to establish the minimum inhibitory concentration or the smallest amount of antibiotic required to stop the growth of bacteria.

Because of penicillin, antibiotic susceptibility testing has become required. Methods like cultivation and dilution, which are phenotypic, were initially used to acquire data. As in the 1980s, polymerase chain reaction (PCR) assays and an antibiotic-impregnated strip called the Etest were available to the public (Choudhary & Tandon, 2009). In addition to developing new tests, researchers are constantly working to improve the accuracy and usefulness of current tests.

2.10: LYSOSOMAL AND PROTEOLYTIC ACTIVITIES

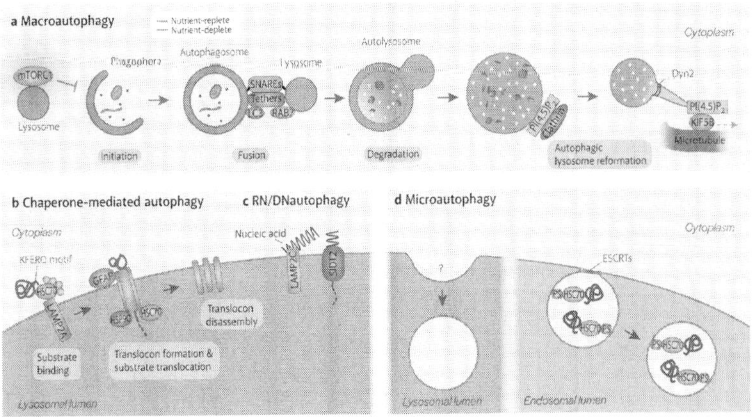

Figure 2.5: Lysosomes.

It is called proteolysis, when proteins are reduced to smaller bits, such as polypeptides or amino acids. An uncatalyzed peptide bond hydrolysis might take centuries to complete. This procedure takes a long time. However, proteases may be found in cells are the most prevalent catalysts for proteolysis. Intramolecular digestion can also produce proteolysis.

Proteins that have been synthesized may need to be degraded by the proteolytic enzymes to become active. Digestive enzymes, for example, break down proteins in a meal so that the body may absorb the amino acids they contain. Additionally, it is critical to maintain the proper folding of proteins and avoid accumulating unwanted or incorrectly folded proteins, in cells in addition to controlling apoptosis and many other physiological and biological processes. Proteolysis regulation deficiency can lead to a various diseases (Chalas et al., 2016).

Aside from being an analytical method in the laboratory, proteolysis may be utilized in industries like food preparation and stain removal to analyze proteins.

Lysosomes are membrane-encased organelles that may be found in almost all mammalian cell types. They are hydrolytic enzyme-loaded spherical vesicles that can degrade a wide range of biomolecules. Because they carry enzymes, they can do this feat. Proteins make up the lysosome's membrane and lumen, and each one is unique in its structure. The pH of the lumen is ideal for the activity of enzymes involved in hydrolysis, similar to how the stomach functions. In addition to its involvement in polymer breakdown, the lysosome plays a crucial role in energy metabolism, secretion, plasma membrane repair, apoptosis, and cell signaling.

Lysosomes play a vital role in breaking down and recycling waste molecules within the cell. The material is first endocytosed through the plasma membrane and placed in a food vacuole. When the food vacuole moves away from the plasma membrane, a lysosome with an active hydrolytic enzyme arrives on the scene. Third-phase hydrolytic enzymes are delivered to the food vacuole by the lysosome, which then combines with it and continues its digestive process. Finally, hydrolytic enzymes are responsible for the breaking down food particles in the fourth and final phases. Lysosomes serve as the cell's garbage disposal mechanism by breaking down worn-out cytoplasmic components inside and outside the cell (Culligan et al., 2009). When the material is brought into the cell from outside, it is called endocytosis. When material already inside the cell is broken down, it is called autophagy. Ten times the size of the smallest organelle is possible. A wide range of organelles can be found in a cell. A Belgian scientist named Christian de Duve was the first to discover and name the phenomenon, which was awarded the Nobel Prize in Physiology or Medicine in 1974.

Lysosomes include around 50 membrane proteins and over 60 enzymes. The CLN6 and CLN8 proteins form a complex that carries lysosomal enzymes to the Golgi apparatus for further processing. Enzyme-carrying vesicles are transported to lysosomes by the Golgi apparatus, where they join more giant, more acidic vesicles. A chemical tag of mannose 6-phosphate is attached to enzymes destined for lysosomes to be more easily sorted into acidified vesicles.

Myofibrillar protein breakdown remains a mystery, even though it has been extensively studied in a various healthy and pathological contexts. Proteases and other hydrolyzing enzymes are found in the lysosome, a cellular organelle now known to contain the lysosome. Because lysosomes do not affect the process of breaking down external macromolecules, one would expect that they play a role in the breakdown of intracellular proteins in muscle cells. Using equilibrium density-based centrifugation procedures, our research facility has discovered that muscle contains two or more distinct populations of lysosomes, each of which can be isolated (Canonico and Bird, 1970; Stauber and Bird, 1974). Actin and myosin are degraded by cathepsins B, D, H, and L. A cytochemical method was used to demonstrate the presence of cathepsins B and D within the muscle lysosomes of mice. (Bird *et al.,* 1978). Sibelius and colleagues discovered in 1979 that secondary lysosomes could contain remnants of myofilaments. Isolated muscles contain radiolabeled proteins, and actin and myosin fragments can be immunoprecipitated after exposure to an actomyosin fragment antibody, according to Gerard and Schneider (1979). The discovery of radiolabeled proteins in isolated muscle lysosomes led to both findings.

A substantial body of evidence suggests that lysosomes in muscle cells are ultimately responsible for the breakdown of myofibrillar proteins in healthy and diseased tissues (Bird, 1975). Finding out exactly how an enzyme contributes to the degradation of myofibrillar proteins in muscle homogenates can be difficult. This is because muscle tissue has a wide range of biological varieties (macrophages, fibroblasts, nerve fragments, epithelial cells, and myoblasts). Therefore, it is difficult to say if myocytes are the only cells capable of producing a specific enzyme. This research found lysosomes and proteases in developing cultured muscle cells, which should help avoid the problem.

In genetically diabetic C57BL/KsJ mice, it has been demonstrated how alterations in cardiac lysosomal and nonlysosomal proteases lead to the development of cardiomyopathy. Mice of the DB genetic background and age-matched control mice were put to death between 7 and 24 weeks. Cathepsin D, myofibrillar alkaline protease (MAP) (including serine protease activity), and Ca2+-activated protease all have their activities measured while using [3H]acetyl-casein as a substrate. In the myocardium of diabetic mice aged 7 to 20 weeks, the activity of cathepsin D, MAP, and serine proteases is significantly decreased. However, by the time the animals reach 24 weeks of age, they have returned to normal levels. Cathepsin D and MAP activity had an inverse relationship to the amount of heat mass seen in diabetic mice, with higher levels associated with a more significant reduction in heart mass (Zhang *et al.*, 2016). Throughout the study, the Ca2+-activated protease activity found in the hearts of diabetic mice did not significantly vary from that found in the hearts of controls. Both lysosomal cathepsin D and nonlysosomal MAP may contribute to the fast degeneration of heart muscle in the late stages of diabetic cardiomyopathy in mice. This is because both of these enzymes are found in lysosomes.

CHAPTER 3

FOODS THAT ARE HIGH IN PROBIOTICS

CONTENTS

Lactis: This potent strain has been proven to boost immune and respiratory system functioning, as well as excellent cell generation and digestive wellness. According to one study, *B. lactis* naturally increases gut lining health by strengthening its integrity and preserving nutrients that should be absorbed in the gut where they belong.

B. longum increases amino acid fermentation, which assists in the upkeep of the body's muscles and general cellular structure. It also ferments carbohydrates, which aids in the preservation of the body's metabolic health. *Lactobacillus acidophilus* and the digestive help it delivers have been examined scientifically. This probiotic strain encourages healthy bowel motions, decreases occasional diarrhea, and helps to maintain bacterial equilibrium in the vaginal ecology. An ongoing investigation looks at how *L. acidophilus* can boost the immune system and promote the body's healthy cell production (Yao *et al.,* 2020).

L. reuteri: BioGaia, a Swedish healthcare center and probiotic research pioneer, has done over 170 clinical experiments on *L. reuteri* alone. *L. reuteri* can minimize infrequent diarrhea and promote overall health by supporting digestive and immune system function when children and adults work together in a compact, crowded situation. It can also aid in maintaining a balanced bacterial population in the vaginal environment.

L. rhamnosus: In addition to avoiding recurring seasonal risks, *L. rhamnosus* may aid in maintaining ideal weight control and acceptable blood sugar levels. By maintaining a healthy balance in the gut flora, *L. rhamnosus*, especially when paired with prebiotics, can help cure occasional indigestion, gas, and bloating.

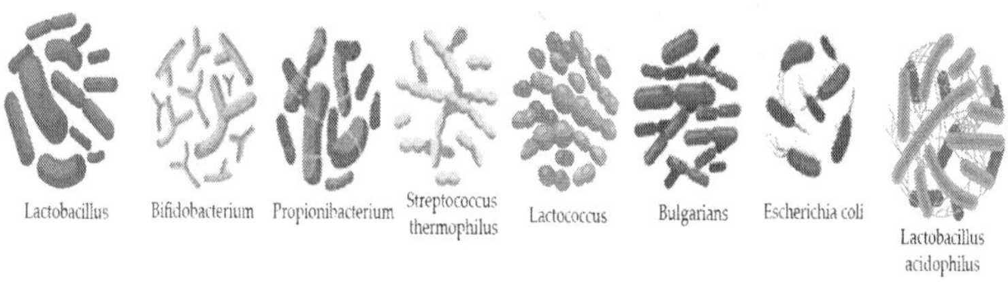

Figure 3.1: Probiotic bacteria.

L. gasseri: It helps the body create folic acid, typically known as folate, which has been proven to promote heart and vascular health. It also combines with other probiotics to boost gut microbial balance and defend against common dietary toxins. *L. gasseri's* exceptional cleansing properties could benefit women in sustaining the balance and health of the vaginal flora.

3.1 WHY ARE PROBIOTICS IMPORTANT?

After deconstructing and evaluating the key strains that may be included in probiotic capsules, it is simple to grasp why they are effective for safeguarding general wellness._They give various other health advantages (Wescombe *et al.,* 2012). Prebiotics are vital components of probiotic supplements that support and promote health in practically every body system.

1. A balanced mix of "good" and "bad" bacteria in your stomach aids optimal digestion and nutrient absorption.
2. Feeding "good" bacteria that may be occasionally overwhelmed by oxidative stress.
3. Promoting proper peristalsis, which turns food into nutrients the body can use for energy, cell development, and repair.
4. Promoting mineral absorption from meals, particularly those that support bone health, such as iron, calcium, magnesium, and zinc (Valero-Cases *et al.*, 2020).
5. Assisting the body's infrequent metabolism of common toxins.
6. Supporting immune cell activity and aiding the body in creating natural antibodies for healthy immunological responses in response to seasonal and environmental challenges.

3.2 WHAT SITUATIONS REQUIRE THE USE OF PROBIOTICS?

Because the human body contains more bacteria than cells, preserving a healthy bacterial balance is crucial for general health. Probiotics can help the overall health regularly, but they can also be helpful sometimes. For example, the doctor may advise you combining probiotics with the antibiotic prescription. Probiotic supplements can also help increase digestive health and reduce indigestion, bloating, and gas. This is especially true following meals.

To assist in obtaining the probiotics the gut craves, add foods like live-cultured yogurt, sauerkraut, miso, kimchi, and kefir to the diet or create a wonderful probiotic smoothie. The diet can also be completed with a high-quality probiotic to strengthen the immune system, digestive system, and general health. Complete Probiotics Platinum® by 1MD Nutrition contains 11 unique strains and offers 51 billion live cultures in each capsule for the best possible support of the digestive and immune systems.

3.3 COMMON FERMENTED FOODS

Food has been preserved by fermentation for thousands of years. This procedure generates wine, cheese, sauerkraut, yogurt, and kombucha. Beneficial microbes are numerous in fermented foods, which have been linked to several health advantages, including increased immunity and digestion (Tadesse, 2012).

3.3.1 What Is Fermentation of Food?

Microorganisms such as yeast and bacteria turn carbohydrates such as starch and sugar into alcohol or acids during fermentation. Due to the alcohol or acids, which also work as a natural preservative, fermented foods exhibit a unique taste and acidity. Additionally, fermentation encourages the formation of probiotics or good microorganisms. Because probiotics have been proven to promote immunological function, digestive, and cardiovascular health, including fermented foods in the diet may improve overall health. During the fermentation process, yeast and bacteria metabolize carbohydrates. It is used to produce yogurt, cheese, and sauerkraut, all of which possess a characteristic sour taste.

3.4 HEALTH ADVANTAGES OF FERMENTED FOODS

The fermentation process delivers many health advantages. Typically, fermented foods have more nutrients than their unfermented equivalents. Below are the critical health advantages of fermented foods:

- Enhancement of Digestive Health

Fermentation results in the development of probiotics that can help restore the appropriate balance of healthy bacteria in the stomach and treat numerous digestive issues. Irritable bowel syndrome (IBS) is a common digestive condition, and probiotics may help improve its unpleasantness Drinking 4.4 ounces (125 g) of yogurt-like fermented milk per day improved symptoms such as bloating and frequent bowel movements in a 6-week study including 274 persons with IBS. Consuming fermented foods may help alleviate the severity of constipation, bloating, gas, and diarrhea (Smolyansky, 2010). For these reasons, incorporating fermented foods into the diet may be beneficial if digestive issues are experiences frequently.

- Immune System Stabilizer

The immune system is greatly influenced by the bacteria in the gut. Due to their high probiotic content, fermented foods can boost the immune system and reduce the risk of illnesses such as the common cold. Furthermore, eating probiotic-rich foods may help recover faster from illness. Furthermore, many fermented foods are abundant in iron, zinc, and vitamin C, all of which have been linked to a more robust immune system that assists in the digestion of food.

- Fermented foodstuffs are easier to digest than no fermented foodstuffs.

Lactose, a naturally occurring sugar in milk, is broken down by fermentation into the simpler sugars glucose and galactose. As a result, lactose intolerant people may consume fermented dairy products like kefir and yogurt. Furthermore, fermentation helps break down and remove antinutrients like phytates and lectins, substances found in seeds, nuts, grains, and legumes that limit nutrient absorption (Stavropoulou & Bezirtzoglou, 2020). Consuming fermented beans or legumes, such as tempeh, improves nutritional absorption and makes them more nutrient-dense than non-fermented alternatives.

- Meals that have been fermented, according to research, may also:
  - Mental health: Several studies have found that consuming the probiotic strains *L. helveticus* and *Bifidobacterium longum* reduces the signs and symptoms of worry and sadness. Fermented foods include both probiotics and prebiotics.
  - Weight loss: Although more research is needed, numerous studies have found links between weight loss and lower belly fat and certain probiotic strains such as *L. rhamnosus* and *L. gasseri*.
  - Heart health: Consuming fermented foods has been linked to a decreased risk of developing heart disease. Probiotics may also help lower total and "bad" LDL cholesterol while lowering blood pressure. Fermented foods have various health benefits, including improved digestive health, higher immunity, and increased availability of essential minerals (Saraf et al., 2010).

3.5 SIDE EFFECTS AND SAFETY

Fermented foods are considered safe by the great majority of individuals. However, some people may have adverse outcomes. Because fermented foods have a high concentration of probiotics, the most common adverse effect is a temporary increase in gas and bloating. These symptoms may exacerbate if fiber-rich fermented foods like kimchi and sauerkraut are consumed. Also, keep in mind that not all fermented foods are the same.

Check nutrition labels to verify purchasing a healthy product, as many foods include high quantities of added sugar, salt, and fat. When fermenting at home, make sure to adhere to the recipes strictly. Food may rot and become unsafe to eat as a result of improper temperatures, extended fermentation periods, or sloppy equipment. Flatulence and bloating may be fermented meals' first adverse side effects (Sistek et al., 2006). Always check the nutrition labels on shop items and follow the instructions when fermenting at home to avoid spoiling.

3.6 FOODS THAT ARE FERMENTED FREQUENTLY

Fermented foods come in a variety of flavors and are famous all around the world, including:

• Kefir • Kohlrabi (kale) • Tempeh The nut • Cheddar cheese (optional) Kombucha is a type of fermented tea • Kimchi is a traditional Korean dish. Snacks like sausages, Greece's Yogurt, a freshly baked loaf of bread; • Beverages containing alcohol • A glass of white or red wine • Olive oil has a high concentration of omega-3 fatty acids.

Fermented foods are popular with customers all around the world. Traditional examples include sourdough bread, cheese, yogurt, kombucha, kefir, and tempeh. When carbohydrates such as starch and sugar are fermented, bacteria break them down into simpler forms. Fermented foods include kimchi, sauerkraut, kefir, tempeh, kombucha, and yogurt. Eating these types of meals can increase immunity, digestion, weight reduction, and lessen the risk of heart disease. To add to the list of fermented foods' advantages, they are a fantastic supplement to a balanced diet.

3.6.1 Yogurt

This meal is made by microorganisms digesting milk. Yogurt culture is a bacteria that is used to manufacture yogurt. Lactic acid, which these bacteria generate, is mixed with milk protein to form yogurt's sour flavor and texture. Cow milk is usually a significant source of yoghurt production (Sidhu et al., 2015). Water buffalo, goats, sheep, mares, camels, and yaks' milk can also be used to make yogurt. Homogenized milk may be used according to liking. Raw and pasteurized varieties are also available. Each type of milk has a unique impact on the body.

Streptococcus thermophilus and *L. delbrueckii subsp. bulgaricus*, two types of bacteria, are cultured to make yogurt. Bifidobacteria and lactobacillus can be added during or after the yogurt culture at any time. Lactobacillus per milliliter in Chinese yogurt must be at least 1 million colony-forming units (CFU).

Milk must be heated at roughly 85 °C to prevent the formation of curds. The milk is allowed to cool to 45 °C after it has been brought to a boil (113°F). Fermentation occurs at temperatures ranging from 30 to 45 °C (86 and 113 °F). Higher temperatures result in faster processing but also increase the possibility of textural lumps or whey separation.

Probiotics, which may be found in yogurt, provide a variety of health advantages, including immune, cardiovascular, and metabolic health. A dearth of high-quality clinical evidence around the turn of the century made it impossible to conclude that yogurt consumption reduces sickness risk or improves health. Postmenopausal women who consumed 80 g of low-fat yogurt daily had a lower risk of developing type 2 diabetes and a lower incidence of hip fracture (Singh *et al*., 2011). Following a thorough analysis, experts found in 2021 that consuming yogurt may assist in maintaining bone health and lessen one's risk of getting diseases like cancer and metabolic syndrome. The study also found that consuming yogurt improved lactose tolerance and digestion.

Safety

Listeria, *Cryptosporidium*, *Campylobacter*, *Brucella*, *E. coli*, and *Salmonella* are among the bacteria in raw milk yogurt that can cause severe sickness or death. Yogurt contains *Aspergillus flavus*, *A. parasiticus*, and *A. nomius*. Although yogurt can be contaminated, industrially processed yogurts are more likely than traditionally prepared yogurts to be infected. Mold cannot be removed from yogurt with a scraper. Because of the thick nature of yogurt, mold grows.

Differentiations and Configurations

- Tzatziki, also known as cack in Greece, is a yogurt-based dip made with cucumbers, olive oil, and fresh mint.

- Dahi is an Indian yogurt recognized for its peculiar flavor and texture. The term dahi, one of the panchamritas, is derived from the Sanskrit phrase dadhi, which translates to "sour milk." Sweetened dahi, also known as mishti doi or meethi dahi, is a popular breakfast and lunch choice in eastern India. Cow milk dominates yogurt manufacturing, although goat and buffalo milk (as in buffalo curd) were popular alternatives due to their high-fat content (Singh & Natraj, 2021).
- Water buffalo milk is used to make Dadih, a traditional fermented bamboo tube yogurt from West Sumatra. Yogurt is a typical dessert and appetizer option in Nepal. Locals describe this cultural staple as "dahi." It is drunk during various events, including religious services and marriage ceremonies. Bhaktapur Yogurt is the Nepalese brand name for yogurt. Yak milk is commonly used to create yogurt, butter, and cheese in Tibet. Yak milk is also known as dri milk since it originates from the male of the animal, the yak.
- Mast Chekide, a sort of kefir yogurt with a characteristic sour flavor, is popular in northern Iran. Delal, a pesto-like blend of fresh herbs and water, is often served alongside it. Popular appetizers include Borani (eggplant or spinach), Mst-o-Khiar (cucumber, spring onions, and herbs), and Mst-Musir (wild shallots). (Foraged shallots) Summer in Iran means a cool soup prepared with yogurt, ice, cucumber, raisins, salt, pepper, and onions and served with Persian bread croutons (Rusu et al., 2019). Lentils and spinach are cooked in a yogurt-based ashe-mast soup. Soups and stews are topped with a sour cream sauce from the water left over after straining yogurt.
- Matsoni, a famous Georgian yogurt, is well-known in Russia and the Caucasus. Tarator and cack, two yogurt-based cold soups, are popular during the summer in Eastern Europe. To make this meal tastier, add garlic, walnut flour, and Ayran. Tzatziki
- Khar w Laban, a cucumber and yogurt salad, is popular in Syria and Lebanon. Furthermore, yogurt is used in various distinctive Lebanese and Syrian recipes, such as "Kibbi bi Laban" Rahmjoghurt, which contains significantly more fat (10%) than most yogurts marketed in English-speaking nations. Dovga, a soup made of yogurt, grains, and herbs, is served hot in the winter

and very cold in the summer (Abuajah *et al.*, 2015). Jordanians enjoy jameed, salted and dried yogurt that has been preserved. Water buffalo milk is used to make Zabadi yogurt, which is popular in Egypt. This beverage is a popular choice for quenching thirst while fasting during Ramadan.

3.6.2 Spiciness and Sweetness

Yogurt is frequently sold sweetened, sweetened, and flavored, or in containers containing fruit or jam to mask its natural acidity. The two most common types of yogurt available in supermarkets are set-style yogurt and Swiss-style yogurt. Swiss-style yogurt is spun before packaging, whereas set yogurt is placed in separate containers. Both beverages can be sweetened by adding fruit (Reid, 2016). The Indian beverage lassi is a swirled mixture of yogurt, salt or honey. The lassi is then flavored with fruit pulp. Unlike rural and rustic lassis, which contain curds or fruit pulp, urban and commercial lassis have a consistent consistency due to processing.

Low-calorie commercial yogurt often contains much sugar or other sweeteners. Certain yogurts are thicker and creamier due to the addition of modified starch, pectin (a natural component of fruits), or gelatin. Although it resembles traditional Swiss yogurt, it can be sold in this manner. Unhomogenized whole milk produces yogurts with a visible layer of cream on top, known as "cream line" yogurts. Single-serving plastic cups are commonly used to sell sweetened and flavored yogurt, which is famous worldwide. Among the most popular flavors are vanilla, honey, toffee, and various fruits. Chocolate and cheesecake-inspired yogurts were popular at the turn of the century. Despite concerns about the potential adverse health effects of sweetened yogurt, a study found that sugar consumption in yogurt manufacturing has decreased since 2016 due to WHO and government efforts to reduce obesity.

3.7 STRAINING

3.7.1 The Method of Straining Yogurt

Various materials, such as paper or cloth, have been used to strain yogurt, including muslin and non-muslin fabric. By omitting whey from the mix, the thickness is increased. Because skim milk has a thinner consistency, straining can be a good way to make homemade yogurt. Labneh is Middle

Eastern yogurt that has been strained to remove the whey. The texture is similar to a cross between yogurt and cheese. It may be used to make sandwiches in Middle Eastern countries (Aggarwal et al., 2013). Optional garnishes include cucumber slices, olives, and other green herbs. Once thickened, formed into balls, and fermented, olive oil can be used to keep it for several weeks longer. It is frequently used to make pies or kibbeh balls with onions, meat, and nuts.

Figure 3.2: Fermented cheeses.

Several varieties of yogurt were traditionally cooked in open vats to reduce the amount of liquid in the yogurt. Eastern yogurts, such as mishti dahi, similar to classic dahi, are sweeter and have a custard-like texture than their western counterparts. In western Indian cuisines, the dessert "shrikhand" is made from yogurt macerated with sugar and spices such as saffron, cardamom, and nutmeg. Tzatziki, a popular side dish for gyros and souvlaki pita sandwiches, is mainly made from strained yogurt. Tzatziki is made by combining Greek yogurt, shredded cucumber, a few drops of extra virgin olive oil, salt, and a minced garlic clove. A delicacy from India, Srikhand is a yogurt-based dessert with saffron and other spices, cardamom, nutmeg, and sugar (Aponte *et al.,* 2020).

Greek yogurt is often called "strain yogurt" because it is so prevalent in the United States To achieve the desired consistency, use powdered milk instead of straining. In the United Kingdom, it is known as "yogurt in the Greek style." In the United Kingdom, only yogurt made in Greece can be labeled as "Greek."

Production

Lactobacillus bulgaricus and *Streptococcus thermophilus*, the bacteria that convert milk to yogurt, must be injected with a starter culture and kept warm for several hours before yogurt can be made by denaturing its proteins (scalding). One option is to boil or keep the milk at 85°C for a few minutes (which produces a slightly different effect). It is frequently chilled to 50 °C (122 °F). The integration of the initial culture and status is the next step.

Yogurt can be made with milk that is higher in solids than regular milk, resulting in a firmer product. To increase the solids content, use dry milk. In the yogurt-making process, heat and acidity, two important inhibitors of pathogen development, are used (low pH). Neither can be overlooked when it comes to producing a safe end product. It has been questioned whether acidity alone is adequate in food poisoning caused by the acid-tolerant *E. coli* O157:H7 strain. Pasteurization (heating) is a simple solution for removing *E. coli* O157:H7 from milk; this first round of heating kills bacteria and denaturized proteins (Qi *et al.*, 2019). Because the bacteria that transform milk into yogurt can withstand higher temperatures than the most pathogens, a suitable temperature not only speeds up the formation of yogurt but also limits the growth of harmful germs. As long as the whey is removed, the yogurt can be thickened and reduced in whey content by filtering after it is made.

3.8 KOMBUCHA

Kombucha (also known as tea mushroom, tea fungus, or Manchurian mushroom) is a lightly effervescent and sweetened fermented black tea beverage. The beverage is sometimes referred to as kombucha tea to distinguish it from the bacteria and yeast culture. Using juice, flavors, fruit, and other substances is standard practice. Kombucha is said to have originated in China, where it is widely consumed.

By the turn of the century, it had spread to Germany, Russia, and other Eastern European countries. Kombucha is made at home, bottled, and sold in stores worldwide. The global kombucha market will be worth US$1.7 billion in 2019. Kombucha is manufactured by fermenting sugared tea with a symbiotic bacterium and yeast culture (SCOBY), sometimes known as a "mother" or "mushroom." A SCOBY can include various bacteria (Prins, 2011). The yeast component is typically Saccharomyces cerevisiae and other species, whereas the bacteria component is typically *Gluconacetobacter xylinus*, which oxidizes the alcohols produced by the yeast to acetic acid (and other acids)

Despite being referred to by the general public as a "tea fungus" or "mushroom," the SCOBY is actually "a symbiotic growth of acetic acid bacteria and osmophilic yeast species in a zoogleal mat (biofilm)." The probiotic advantages of living bacteria are one of the reasons for the drink's popularity. Various health benefits have been ascribed to kombucha use, although there is little evidence to support them. Contamination during home preparation might have caused the beverage to have rare but severe side effects. It is not advised for therapeutic purposes. Kombucha can be brewed at home or in a commercial setting.

Kombucha is made by dissolving sugar in boiling, non-chlorinated water. The tea leaves are evacuated after steeping in hot sugar water. After the sweetened tea has cooled, the SCOBY culture is added. The slurry is placed in a sterile beaker with previously fermented kombucha tea in a procedure known as "backslopping." [The container is covered with a paper towel or permeable fabric to keep pests like fruit flies from contaminating the kombucha. The tea can be fermented for 10 to 14 days at room temperature (18 to 26 °C). A new "daughter" SCOBY will develop on the tea's surface until it reaches the size of the container (Pratap *et al.,* 2020). When the fermentation is complete, a small amount of the freshly fermented tea is stored with the SCOBY. The remaining kombucha is filtered, bottled, and stored at 4 °C for a brief secondary fermentation.

Kombucha was first commercially available in bottles in the late 1990s. In 2010, high alcohol levels in various bottled kombucha products were discovered, prompting Whole Foods and other retailers to remove the beverages off their shelves. As a result, kombucha brewers altered the alcohol content of their products. As of 2021, the beverage was popular in India's National Capital Region, thanks partly to its popularity in the West. In India, kombucha producers include Bhu Kombucha, Stoked Kombucha, and Toyo Kombucha. In 2019, "hard kombucha" with an alcohol concentration of more than 5% was discovered.

Figure 3.3: Homemade raw kombucha.

3.8.1 Claims Relating to Health

Kombucha tea has grown in popularity due to its purported health benefits. Among other things, the products are said to cure AIDS, aging, anorexia nervosa, arthritis, atherosclerosis, cancer, constipation, and diabetes. Its possible biological effects on humans have not been studied, and the potential health benefits associated with its biological activities have not been demonstrated. In a 2003 systematic review, kombucha was singled out as an "extreme example" of an alternative therapy due to the mismatch between the product's apparent hazards and its implausible, broad-based benefit claims (Pandey *et al.,* 2015). It ruled that the proposal's unproven therapeutic claims outweighed the known risks and that kombucha should not be recommended for therapeutic use because it falls into the category of "remedies that appear to benefit mostly those who sell them."

3.8.2 Negative Ramifications

According to the 2003 study, adverse effects from kombucha consumption are uncommon but may be underreported. Consuming Kombucha tea has been linked to serious negative outcomes and a few fatalities. Because kombucha is frequently made at home, caution should be exercised because potentially harmful microorganisms may contaminate the tea during preparation. Consumption of kombucha has been linked to severe liver and kidney damage and metabolic acidosis.

Although the beverage has never been definitively proven to cause death, it has been linked to at least one fatality. Brewers are encouraged to avoid over-fermentation because the acidity of the tea, which can cause acidosis, can have serious health consequences. Bacterial or fungal infections can have serious consequences throughout the brewing process. Many studies have found usnic acid, a hepatotoxin, in kombucha; however, it is unclear whether this molecule or another toxin is to blame for the liver damage.

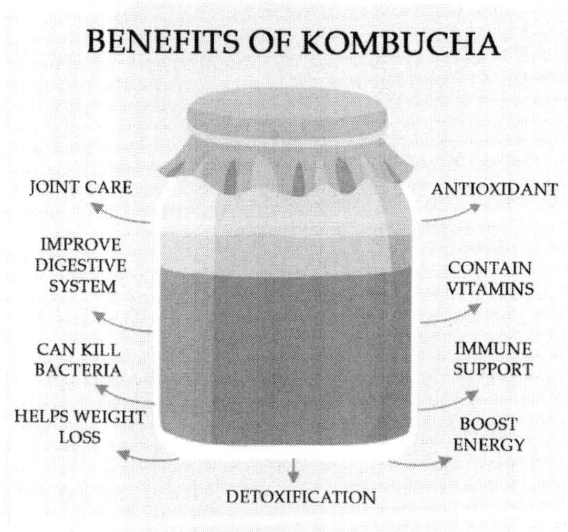

Figure 3.4: Benefits of kombucha.

Consumption of Kombucha may be hazardous to those who already have health concerns.

Those with compromised immune systems, pregnant or nursing women, and children under four should avoid kombucha due to its microbial origin and potentially non-sterile packaging. In susceptible individuals, it may cause digestive or immune system damage.

Because kombucha does not contain alcohol, it should not be mixed with other beverages. A 2019 systematic review concluded that "kombucha is not considered harmful if consumed in amounts of about 4 oz [120 mL] per day by healthy individuals; potential risks are associated with a low ph. brew leaching heavy metals from containers, excessive consumption of highly acidic kombucha, or consumption by individuals with pre-existing health conditions." (Attri *et al.,* 2021).

Additional applications

When kombucha culture is dried, it transforms into microbial cellulose, a leather-like material that can be shaped into shapes to create seamless garments. Although many plant-based dyes can be used to color the cloth, the kombucha culture produces a wide range of textile colors when grown on various broth mediums, such as coffee, black tea, and green tea. The feel and texture of the cloth may differ depending on the growth medium and colors used. The dried SCOBY can be eaten as a savory or sweet snack.

3.9 SAUERKRAUT

Fresh, thinly sliced cabbage has been treated with lactic acid bacteria. Lactic acid is produced by bacteria that break down carbohydrates in cabbage leaves, which adds to the vegetable's extended shelf life and sour fragrance. It is one of Germany's most well-known dishes. Many people are familiar with its German name, but it is well-known throughout Eastern Europe and beyond.

Benefits

Sauerkraut is considered to provide several health advantages, including:

- Because of the fermenting process, sauerkraut is a more nutrient-dense food than cabbage. Vitamins C and K are present in appropriate concentrations. This dish contains vitamins and minerals, including folate, iron, and potassium. It has a low-calorie count but a high calcium and magnesium content.

Figure 3.5: Steamed sauerkraut.

- Raw and unpasteurized sauerkraut includes enzymes, live lactobacilli, and healthy bacteria. The effects of fiber and probiotics on digestion and the creation of healthy gut flora make it possible to prevent various digestive system disorders. This series includes numbers 28 and 29. Canker sores can be treated with sauerkraut juice, a well-known traditional treatment (Parvez et al., 2006). Gargle with sauerkraut juice for 30 seconds every day, or apply a wad of sauerkraut to the afflicted area and hold it there for a minute before swallowing.

Disadvantages

Excessive consumption of sauerkraut may cause bloating and gas due to the difficulties of the small intestine in breaking down the trisaccharide raffinose. Even though it is unpleasant, there are no long-term health consequences. Sauerkraut also contains a significant amount of salt.

3.10 PICKLES

Pickles may enhance the flavor of a favorite burger or sandwich in more ways than one. Cucumber pickled in a sour brine is high in vitamins and minerals. People began pickling them around 4,000 years ago to preserve and extend their shelf life. It is now possible to pickle cucumbers at home. Stores sell whole dill pickles, sliced sweet pickles, and sour spears.

Dietary Habits

Pickles contain no fat or protein and are almost entirely water. Pickles are a good source of vitamins because saline brine extracts water from them. Pickles are made from natural ingredients. Fermented foods like kimchi, kefir, and miso benefit the gut by aiding digestion (Novik & Savich, 2020). Pickles sold in supermarkets are not fermented, a food preservation process that involves yeast, bacteria, and other microorganisms. Pickles are soaked in vinegar and spice brine to give them a distinct flavor.

Advantages to Health

- Reduces digestion time: Probiotics, which are necessary for a healthy digestive system, can be found in fermented pickles.
- Combats illness: Cucumbers contain beta-carotene, an antioxidant that the body converts to vitamin A. Carotene has been shown

to reduce the risk of death from diseases such as heart disease, stroke, cancer, and respiratory problems.

- It May be beneficial in reducing muscular tension: Pickle juice is recommended by several professional athletes as an electrolyte source. According to one study, pickle juice may be slightly more effective than water in relieving muscle cramps. However, there is no conclusive evidence (Nichols, 2007).

Lowering sugar peaks is required. Pickle juice, particularly vinegar, may aid in blood sugar stabilization. Those who are at risk of developing diabetes may benefit from this.

Keep an Eye Out for These Symptoms

Pickles' saltiness is a significant issue. One big dill pickle has more than two-thirds of an average adult's recommended daily salt intake. A high-salt diet can aggravate hypertension and other cardiovascular and kidney diseases. Furthermore, salt can dissolve calcium in the bones. As a result of this, one may be more prone to fractures.

3.11 KIMCHI

It is easy to understand why kimchi, a mainstay in Korean households for decades, has been elevated to kitchen rock star status. Kimchi's appeal is broad and deep due to its unique flavor and versatility. It is a sweet, sour, and spicy mixture of veggies, garlic, ginger, and fish sauce that can be served on its own as a condiment, ingredient, dip, or side dish. The profile includes sour, spicy, and/or sweet tastes. Probiotics, which give nutrition information, are found in fermented meals.

3.11.1 What Is Kimchi, Exactly?

Kimchi is a classic Korean meal prepared with various vegetables, garlic, ginger, chili peppers, salt, and fish sauce; however, the specific components may vary. The veggies are pickled and fermented, a centuries-old way of storing them for the winter. Although carrots, radish, cucumber, and scallions are all often used to prepare kimchi, cabbage is the most usually utilized vegetable. Various kimchi recipes may turn vegan by using plant-based components (Nami *et al.*, 2015).

Commercial kimchi is widely available and moderately priced, but you may opt to create your own depending on your tastes. Preparing the veggies, like other fermented foods like bread and kombucha, may take time and effort, but it is usually a hands-off experience. It will establish its taste and nutritional profile throughout this operation. Kimchi may be stored in the refrigerator for an extended period, making it an affordable, flexible, and simple-to-prepare food item.

Figure 3.6: Kimchi.

3.11.2 Kimchi Applications

Kimchi is served with every meal in Korean culture, even morning. Kimchi is typically served as a side dish or appetizer, but it may also be utilized in several cuisines such as Japanese, Korean, and Mexican. The traditional Korean meal Kimchi jjigae is the most popular way for many Koreans to consume Kimchi. It is also used to season fried rice, stir-fry, noodles, sandwiches, and pizza. Because kimchi is such a diverse meal, the flavor changes significantly from recipe to recipe (McKenzie *et al.*, 2018).

The key flavor qualities of kimchi are sour, spicy, and umami. Aside from the veggies used, the fermentation period and salt/sugar concentration will impact the flavor. Kimchi's characteristic flavor is generally sour due to fermentation. Lactic acid, created by bacteria during fermentation, gives food an acidic, pungent flavor comparable to sauerkraut. When fermented, garlic adds taste to kimchi. Kimchi may be hot if a lot of pepper is used, but it can also be umami-rich if a lot of fish paste, fish sauce, or anchovies is used. Kimchi made with radishes or cucumbers instead of fish has a more delicate taste.

3.11.3 Recipes

Kimchi lends itself so effortlessly to experimenting that the possibilities are virtually limitless. This meal gives tofu, a culinary blank slate that absorbs the tastes of whatever is cooked with it, a gorgeous texture and flavor counterpoint. In addition to kimchi-pickled eggs and savory pancakes, sriracha sauce can be drizzled over it for a small amount of heat (Mahasneh & Abbas, 2010). It is so flexible that it can be used for almost every cooking chore can be thought of. To prepare it as vegetarian, leave out the fish-related components or look for recipes that do.

3.11.4 Where to Buy Kimchi

Kimchi is slowly gaining popularity worldwide, and it is now available in many stores. Kimchi is commonly found in refrigerated vegetables, pickles, and sauerkraut. Kimchi is available at Asian supermarkets, restaurants, and sushi bars. Many restaurants and other businesses manufacture their kimchi and serve it as a side dish. Making kimchi at home is a straightforward technique that requires only a few materials and a few days of fermenting.

Storage

Kimchi may be stored in the refrigerator for several months. However, the flavor will be enhanced, and the veggies will lose some crispness.

3.12 SOURDOUGH BREAD

Sourdough bread appears to be gaining popularity again as hundreds of individuals are forming their beginnings and developing handmade recipes. Sourdough bread has existed for thousands of years, although cooking it has become a popular pastime. It is the first known sort of leavened bread (bread that rises as a consequence of yeast or other components), and several civilizations worldwide enjoy it (Li *et al.*, 2016).

This is tied to how simple it is to prepare sourdough bread. Sourdough bread relies on a starter, a combination of water and wheat, that generates a colony of natural yeast, rather than using baker's yeast. The lactic acid created by this yeast offers sourdough bread its uniquely tangy flavor. This acid also keeps a sourdough starter from degrading by flavoring the bread and eradicating undesirable microbes.

Today, sourdough bread may be cooked at home or purchased at bakeries and supermarkets around the nation. While sourdough bread is generally touted for its health advantages, not everyone enjoys it. Here is a list of the advantages and disadvantages of sourdough bread. Antioxidants are also included in sourdough bread. According to research, antioxidants like the peptides in sourdough might lower the risk of having certain cancers, aging-related symptoms, or chronic illnesses like rheumatoid arthritis.

3.12.1 Benefits of Sourdough Bread for Health

An excellent source of vitamins and minerals is sourdough bread. However, sourdough bread's health benefits might make it problematic for individuals with particular medical concerns. There are numerous probable health advantages of eating sourdough bread, according to research:

- Basic physiological processes

Particularly rich in nutrients that the body may easily absorb is sourdough bread. This is a result of the interactions between these nutrients and the lactic acid bacteria in the bread. These bacteria eliminate certain kinds of an acid generally found in other sorts of bread, improving the availability of vitamins and minerals, including folate, potassium, and magnesium. The body requires folate to divide cells and generate DNA and other genetic material. Additionally, potassium supports the operation of the cells. It maintains the heart rate under control, supports appropriate neuron and muscle activity, and is essential synthesizing proteins and breaking down carbohydrates (Lin *et al.*, 2019). The blood sugar and protein levels are handled, your muscle and neuron activity is controlled, and magnesium aids in the production of protein, bones, and DNA.

- Health Promotion

Additionally, sourdough bread delivers more antioxidants than other forms of bread due to the lactic acid bacteria. The antioxidants defend the cells against cellular damage that may result in horrible disorders, including cancer, heart disease, Alzheimer's disease, and others.

- Decreased amounts of blood sugar

Compared to white bread, sourdough bread may help you maintain more stable blood sugar levels. The sourdough-producing bacteria also have a notable impact on the bread's starch. The bread's glycemic index reduces due to the modified bread molecule structure, which helps the body absorb carbs more slowly. This suggests that compared to eating white bread, a slice of sourdough bread will not cause the insulin levels to rocket as high.

- Enhancing Digestive Health

For some people, sourdough bread may be simpler to digest than white bread. Sourdough bread serves as a prebiotic, which means that the fiber in the bread helps feed the "good" bacteria in your intestines, in line with an expert study. The stability and health of the digestive tract depend on these microorganisms (Kumar, 2013).

Additionally, sourdough has less gluten than other varieties of bread. It appears like the bread's acid is removing gluten. As a consequence, sourdough may be gentler on the stomach for individuals who are gluten intolerant.

3.12.2 Risks Associated with Eating Sourdough Bread

Even if sourdough bread is advantageous, there are still significant hazards linked. Before consuming sourdough bread, think about the following:

- Contamination

The method of creating sourdough at home is popular. While this is typically totally safe, a tainted sourdough starter might emerge. The starter for sourdough might be poisoned if it:

- Does not bubble
- Forms patches that are green, pink, orange, or black.
- Looks "fuzzy."

If the sourdough starter demonstrates any of these traits, discard it.

- Not free of gluten

Despite often containing less gluten than other bread, sourdough is not gluten-free. Sourdough bread may be simpler to digest for persons sensitive to gluten; however, celiac disease patients may probably have symptoms after ingesting it (Kukkupuni *et al.*, 2015). Sourdough bread should not be included in the diet in case of celiac disease or gluten sensitivity, so check with a doctor first.

3.13 CHEESE

Cheese is one of the most popular dairy products in the world. There are numerous kinds of cheese to choose from. Hard cheeses like Parmesan or grana Padano, soft cheeses like mozzarella or burrata, and soft-ripened cheeses like Camembert are all delicious. Popular cheeses like Cheddar, Gouda, and Colby are on the list and are eaten by millions of people every day (Kim & Park, 2017). There are numerous methods for processing and transforming milk into cheese, each with its distinct flavor, texture, and appearance. Some cheeses have a mild butter flavor, while others have a nutty, robust aroma.

3.13.1 Cheese Species

There are hundreds of different types of cheese, but they can be divided into six primary groups:

- Cheese that is not overly squishy. In terms of popularity, Parmesan is the most widely available hard cheese, followed by Pecorino and Grana Padano. Semi-hard (firm) cheese is an option (semi-hard cheese). Cheddar, Gouda, Provolone, and Marble are these types of cheese.

Blue mold is found in cheeses such as Roquefort, Gorgonzola, Stilton, and Danish Blue. The cheese has melted and become pliable. This category includes cheeses such as Brie, Camembert, Mozzarella, Burrata, and Feta. Cream cheese, cottage cheese, and queso fresco are fresh cheeses that can be served at room temperature or chilled (Khoruts, 2018).

Cheddar cheese contains a lot of calcium. Nothing beats cheese when it comes to food. Despite its deliciousness, many people avoid cheese because it is high in saturated fat. When selecting a favorite cheese, keep in mind that there are healthier options to consider. Cheese is also high in protein, calcium, and other nutrients.

3.14 PRESERVING PROBIOTIC FOODS

Let us examine more closely the importance of probiotic survival. Because probiotics originated in the dairy sector, a link between the two appears unavoidable. This arrangement and a love marriage have certain similarities. Dairy chillers are good places for probiotic bacteria to dwell since they thrive at colder temperatures. Probiotics may now be added to airy and non-dairy products to provide unique food and beverage offers.

In other circumstances, refrigeration may not be required. Bifidobacterium and Lactobacillus, two early probiotic bacteria, were used to fortify meals and beverages. Both of these strains must be maintained refrigerated to avoid early germination. These two strains should not be used to make products that must be frozen, heated, or shelf-stable or that must withstand rigorous production techniques (Jugan et al., 2017). Probiotics that generate spores, such as BC30TM, may usually survive in most circumstances without refrigeration (where there is a water activity of less than 0.7). To do this, they must remain still until they reach the digestive tract.

3.15 PROBIOTICS WITH ANTI-INFLAMMATORY PROPERTIES

BC30 allows businesses to employ probiotics in ways other than typical dairy applications to stimulate innovative product development. Regardless of recent advances in probiotic technology and research, assessing whether a strain can endure the production procedures necessary for an application necessitates the analysis of strain and quality data.

CHAPTER 4

POTENTIAL BENEFITS OF PROBIOTIC PRODUCTS

CONTENTS

Probiotics are naturally occurring and beneficial bacteria and/or yeast. Germs have a negative connotation because they are commonly associated with illness. However, both good and bad bacteria always exist in the body. Probiotics contain beneficial bacteria that help the body stay healthy and function properly by boosting the immune system. When the body has too many dangerous bacteria, these beneficial bacteria can help in various ways, including feeling better (Johnston *et al.,* 2012).

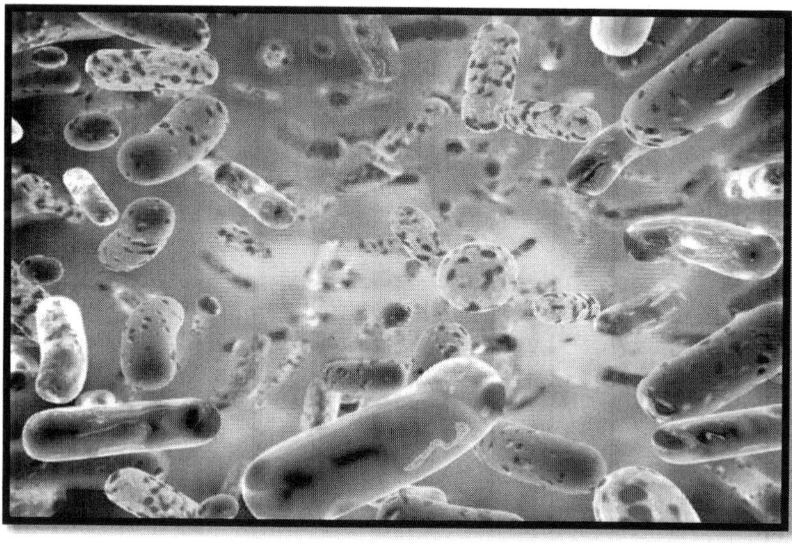

Figure 4.1: An illustration of *Bifidobacterium*, a bacterium.

Probiotics, for example, are part of the microbiome, including all of the body's bacteria. When it comes to keeping the body healthy, the microbiome is like a forest of interconnected creatures. These microbes are referred to as "community members." Health is derived from the millions of microorganisms that populate the body (Hill, 2010). A wide range of microorganisms is included in these terms.

4.1 YEASTS (AS WELL AS OTHER FUNGI)

Everyone's microbiome is unique. Microbial cells can differentiate between identical twins. There are no two people who are alike. To be considered a probiotic bacterium, it must have a diverse set of properties (Harzallah & Belhadj, 2013). Some examples are as follows:

- Quit any social situations that could be endangering.
- After consumption, they survive in the digestive tract (being eaten).
- Has proven to be useful.
- Consumption is permissible.

Where can I find my body's beneficial bacteria (probiotics)?

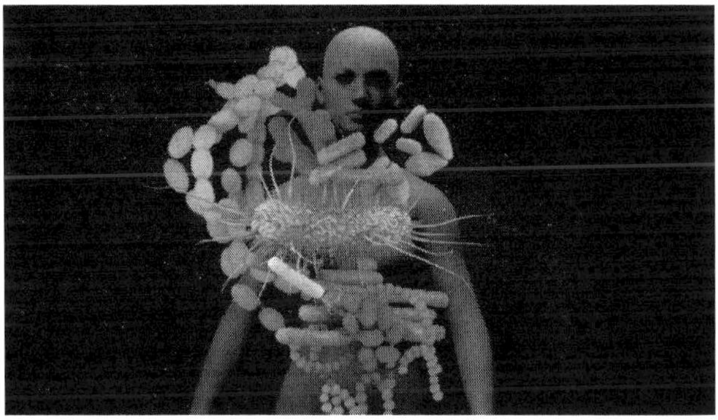

Figure 4.1: The human microbiome.

Although beneficial microorganisms are most commonly associated with the stomach, they can also be found in other parts of the body (mainly the large intestines). These locations have access to the "external world." This is a heartbreaking experience.

- Ears and a nose
- In the genital sense
- The urogenital tract
- The skin is the first thing that must be considered.
- The respiratory system

Probiotics work unusually. The primary goal of probiotics, also known as "good" bacteria, is to maintain a healthy microbiome. Consider acting in a state of physical equilibrium. Bacteria enter the body and multiply while a person is ill. As a result, the body is thrown off. Good bacteria fight off bad bacteria and restore body balance, improving the condition (Gwee *et al*., 2018). The primary advantages of beneficial microorganisms maintaining a healthy immune system and reducing inflammation. Be capable of promoting the growth of beneficial microbes.

4.2 KEEPING BAD BACTERIA FROM MULTIPLYING

According to Harvard Medical School, the intestines are home to 100 trillion bacteria representing over 500 different species. Each person's germ mix is as distinct as their fingerprints. Illness or stress can disrupt the bacterial balance, resulting in digestive and other issues. Probiotics affect existing gut bacteria by altering their genetic makeup or metabolic activity. Beneficial bacteria outnumber harmful bacteria in your intestines. This method prevents pathogenic bacteria from causing illness or inflammation. A yeast infection, for example, can be caused by an overabundance of yeast in the body, but a healthy gut biome keeps yeast levels low.

Probiotics may benefit digestive health and nutrition absorption. Enzymes or proteins produced by beneficial bacteria can suppress or eliminate pathogenic germs. Taking specific probiotic species can help to boost the immune system. Specific microorganisms are required for hormone production, vitamin absorption (such as vitamin K), and nutrition absorption (Gogineni *et al.*, 2013). Bacterial deficiency in children may be linked to the development of type 1 diabetes. Biome data that is trustworthy.

Probiotics can be used for a variety of purposes, including:

- Assists in digestion and prevents diarrhea.
- Prevent genital infections
- Lowers chances of developing autoimmune diseases.
- Treats skin problems.
- Averts urinary tract infections.

However, remember that not every probiotic will have the same effect on everyone's body. Some doctors recommend probiotics in addition to antibiotics when treating a specific illness. Because antibiotics kill both good and bad bacteria, they can cause diarrhea. Probiotics can help to maintain or restore intestinal homeostasis. There are thousands of different types of bacteria and germs in the body. Various microorganisms live in the intestine, also known as the gut (Gupta, 2011). The microbes in the gut are highly interconnected. Bacteria may be beneficial or harmful to human health if the intestinal environment is hostile to them. It is possible to be infected with dangerous bacteria while displaying no symptoms. However, if the right conditions exist, an infection can develop.

Staphylococcus aureus, for example, can be found in the nose and intestines without causing harm. Skin infections are among them, as are more sever conditions such as pneumonia and sepsis. An antibiotic-resistant strain of the bacteria, Methicillin-resistant *S. aureus* (MRSA), is currently infecting hospital patients. Otherwise, healthy patients can become infected with MRSA.

Bacterial infections that cannot be treated pose the greatest risk. Probiotics, or beneficial microbes, are being researched as a possible way to prevent S. aureus infections. Bacillus bacteria, found on soil-grown vegetables, are used in various probiotic formulations. To survive the harsh conditions of the digestive tract, they produce spores that can withstand them(Gao *et al.*, 2022).

The director of the National Institute of Allergy and Infectious Diseases, Dr. Anthony S. Fauci, explains why probiotics are commonly prescribed as digestive health supplements. Several studies have shown that they can aid in the improvement of health. Oral bacteria may be a better therapy option for certain infections than antibiotics, but more research is needed. "

4.3 CREATES VITAMINS. HELP SUPPORT THE CELLS THAT LINE THE GUT TO PREVENT BAD BACTERIA FROM ENTERING THE BLOOD

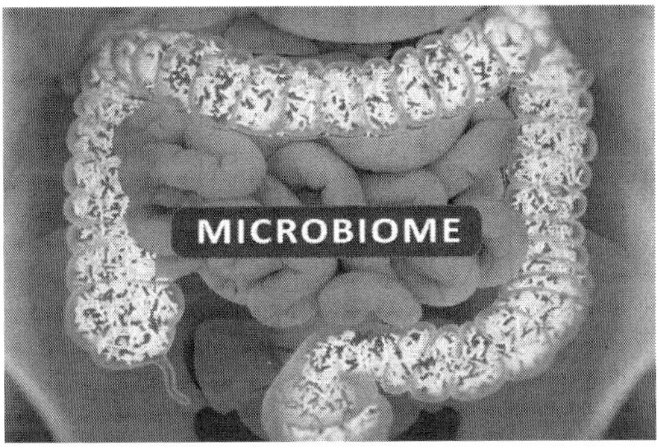

Figure 4.2: Gut microbiome.

A recent study of the human gut microbiota shows that differences in microbial populations may increase vulnerability to various disease presentations. Intestinal bacteria can convert dietary nutrients into metabolites, which act as physiologically active compounds that alter the host's regulatory systems (Faujdar *et al.*, 2016). Probiotics may alleviate or prevent gut inflammation and other intestinal or systemic disease symptoms by restoring the composition of the gut microbiome and providing positive functionalities to gut microbial populations.

4.4 THE IMPACT OF THE GUT MICROBIOME ON HUMAN HEALTH

Joshua Lederberg introduced the concept of the human microbiome to the scientific community, defining it as "the ecological community of commensal, symbiotic, and pathogenic microorganisms that share our body space and have been all but ignored as determinants of health and disease." Since the formation of the Human Microbiome Project (HMP), studies indicating the microbiota composition in healthy and ill human populations and cutting-edge data mining techniques and data comparison analyses, have been published. An investigation into the faecal microbial communities of adult twin pairs who were both skinny and overweight yielded an unexpected result. In contrast to microbial makeup, the findings indicated that each individual's human gut microbiome might have a recognized core set of genes and pathways. Obesity has also been linked to changes in the microbiota at the phylum level and changes in the proportion of bacterial genes and metabolic pathways. Several significant microbial biomarkers of obesity involved in glucose, lipid, and amino acid metabolism were discovered through metagenomic analysis (Ezema, 2013). Changes in core microbial functions, rather than core microbial communities, may be associated with physiological or pathological changes.

One of the methodologies used in comparative metagenomics is extensive searches against databases containing information on functional cellular networks. The Kyoto Encyclopedia of Genes and Genomes (KEGG) and Clusters of Orthologous Groups (COG) analysis are two examples of such databases. These studies discovered that specific genetic components of the intestinal microbiota might functionally complement human genes that are missing or poorly encoded in the human genome but are important for major biological pathways in the intestine. Genes encoding xenobiotic detoxification enzymes, methanogenic pathways for hydrogen gas removal, and polysaccharide metabolism activities have all been identified as enriched functional genetic categories within gut microbial populations.

Intestinal microorganisms that modify gene expression in the mammalian gut mucosa can alter the function of the gastrointestinal system. According to a study using germ-free and conventionally bred mice, the gut microbiota altered the expression of many genes in the human or mouse intestinal tract, including genes involved in immunity, nutrition absorption, energy metabolism, and intestinal barrier function. It is worth noting that the small intestine's mucosa changed the most (Deng *et al.*, 2020). Probiotics in the digestive tract have the potential to alter gene expression.

4.5 THE IMPACT OF DIET ON GUT FLORA

Scientists have debated whether nutrition affects gut flora since the 1960s. Recent studies have used animal models, intestinal microbiota investigation, and metagenome analysis to investigate the relationship between nutrition and the makeup and function of the gut microbiome. Human meals may influence the microbiome by altering biochemical reaction patterns in the gut lumen. Animals fed a high-fat, high-sugar Western diet showed rapid changes in the structure of the intestinal microbial community, with an increase in the number of members of the phylum Firmicutes and a decrease in the abundance of members of the phylum Bacteroidetes in experiments using germ-free mice transplanted with human fecal microbiota. Genes involved in glucose transport were overrepresented in metagenomic studies (Deshpande *et al.*, 2011).

However, after a week of resuming a typical chow diet, microbial populations returned to normal. This theory underpins the concept of "luminal conversion." Microorganisms in the intestine absorb and convert nutrients taken in by the host into other molecules, such as vitamins, amino acids, or dietary fiber. SCFAs, biogenic amines (such as histamine), and other amino-acid-derived metabolites such as serotonin and gamma-aminobutyric acid (GABA) may be physiologically active in healthy and diseased conditions. The production of these compounds may have an impact on microbial composition changes. Nondigestible carbohydrates from food can ferment in the intestinal lumen, producing SCFAs like acetate, propionate, and butyrate. Human colonic epithelial cells use metabolically active SCFAs involved in various biological functions as a source of energy. Furthermore, the fermentation of prebiotic carbohydrates like inulin and fructo-oligosaccharides promotes the growth of beneficial microorganisms like *Bifidobacterium* and *Lactobacillus*.

There is new information about the composition and function of healthy gut flora. The stools of 242 healthy young people were examined using complete metagenomic sequencing and 16S rRNA gene pyrosequencing to determine the content and function of the microbiome [The Human Microbiome Project Consortium]. In terms of microbial taxa and genes, the distal intestine's biological diversity and richness far outnumber other body locations such as the mouth cavity or human skin. The classification of organisms, as expected, Bacteroidetes and Firmicutes, predominated in the human intestine. The major taxa varied throughout the body.

Different bacterial genera and species have varying degrees of dominance. *Bacteroides fragilis*, for example, was found in at least 0.1 % of sequencing reads in 16 per cent of distinct individual specimens (Di Cerbo & Palmieri, 2015). Furthermore, *Bacteroides thetaiotaomicron* was found in 46 % of participants, accounting for at least 0.1 % of sequence reads. Bacteroides taxa are well-known commensal intestine taxa cultured and studied in the laboratory for their immunomodulatory and metabolic abilities.

In contrast to microbial composition, whole genome metagenomics data revealed that metabolic pathways were evenly distributed and shared across body regions and individuals. Dominant metabolic modules, such as central glucose metabolism, reflect critical functional categories in the distal intestine and other body parts. Despite this, 86% of the discovered gene families from gut metagenomes have yet to be functionally defined or fully mapped to pathways.

4.6. THE ORAL PHARMACOKINETICS OF ACETAMINOPHEN IN MICE AS A RESULT OF PROBIOTIC SUPPLEMENTATION

Oral probiotics influence the microbiota and enzyme activity. The co-administration of probiotics and medications can potentially change the medications' pharmacokinetic properties. This study looked at the pharmacokinetics of acetaminophen in mice given probiotics. Oral probiotic treatment altered the composition of mice's gut microbiota. According to the study, consuming probiotics may disrupt drug metabolism by affecting the gut microbiota's ability to metabolize pharmaceuticals, resulting in a change in drug concentrations in the body (Daliri & Lee, 2015).

Oral medications enter the bloodstream via the gastrointestinal (GI) tract. Permeability, stability, solubility, and metabolism by the body's enzymes and microorganisms in the intestines are all factors that influence how well a drug is absorbed. Although this has not been thoroughly studied, gut bacteria may influence the rate at which medications are excreted into the blood. Xenobiotics, including bacteria in the intestines, can be metabolized by any organ in the body, including the liver. The liver has long been thought to play an essential role in drug metabolism. According to recent research, ingested xenobiotics undergo an initial metabolism in the gut flora before entering the bloodstream. The liver produces hydrophilic metabolites via oxidative and conjugative reactions such as glucuronidation and sulfation in contrast, hydrophobic metabolites are produced by the gut microbiota via reductive and hydrolytic reactions. As a result, the metabolic activities of gut bacteria differ markedly from those of the liver. The former is a detoxification reaction, whereas the latter is a toxification or activation response.

The most commonly prescribed analgesic and antipyretic is acetaminophen. Acetaminophen oral ingestion activates the phase II enzymes arylsulfotransferase and UDP-glucuronyl transferase, which catalyze glucuronidation and sulfation. By detoxifying the drug in the intestinal mucosa after oral administration, these enzymes aid in the elimination of acetaminophen from the body via the bile and urine. When acetaminophen is metabolized in small amounts, N-acetyl-p-benzoquinoneimine is formed and conjugated with glutathione. When acetaminophen is taken orally, acetaminophen-sulfate and cresol-sulfate compete for excretion in the urine (Dai *et al*., 2018). Acetaminophen-sulfate excretion in the urine is increased by intravenous administration of p-nitrophenyl sulfate. It is found in the gut bacteria *Eubacterium rectale* A-44 and *Enterobacter amnigenus* AR and deactivates the enzyme that sulfates acetaminophen, cresol, and tyrosine. Changes in gut flora can affect acetaminophen pharmacokinetics both internally and externally.

Yoghurt and kimchi are popular probiotic sources due to their beneficial physiological effects on constipation, diarrhea, colitis, and diabetes, as well as their ability to prevent the production of harmful enzymes by the human gut microbiota, which is why they are considered beneficial bacteria. As a result, the use of probiotics in therapeutic settings is increasing. Pharmaceutical medications and probiotics can be used together to treat certain disorders. Specific probiotics can influence the pharmacokinetic properties of orally administered medications, such as nifedipine metabolism via CYP3A in the intestinal mucosa or sulfasalazine metabolism by gut flora.

4.7 IMPROVEMENT OF INTESTINAL HEALTH BY THE REGULATION OF MICROBIOTA

Probiotics, which are beneficial microbes, can alter intestinal flora. Probiotics may help to maintain the intestinal epithelial barrier, prevent pathogen adhesion to the gut surface, and regulate and mature the immune system. Furthermore, probiotics have been shown to treat various conditions by altering the microbiome and boosting the host's immune system. According to the most recent research, probiotics, gut flora, and immunity are all intertwined. Probiotics will be investigated further to learn how they affect gut flora composition, immunity, and human well-being (Chugh & Kamal-Eldin, 2020).

The gut microbiota has received much attention because of its apparent link to the host's overall health. Disruption in the stomach's normal functioning can result various health issues. If the immune response of the central nervous system is compromised due to changes in the gut microbiome, this could lead to mental health issues such as depression, Parkinson's disease, and schizophrenia. The intestinal microbiota influences the development of the central nervous system via the gut-brain axis. A disruption in the gut-liver axis can result in a potentially dangerous bacterial population. The risk of liver damage grows as the toxic compounds accumulate. Changes in the gut microbiota can cause food allergies, diarrhea, and even intestinal cancer. As a result, the human gut microbiome and health must be closely monitored. To control the gut microbiota, functional diets now include probiotic and prebiotic supplements.

There is compelling evidence that probiotics and prebiotics improve human health and wellness. The ability of probiotics to improve a person's health by balancing the microbiota in the digestive tract is well-established. To promote the growth and activity of healthy gut flora, prebiotics can be used in addition to or instead of probiotics (GIT). Naturally occurring prebiotics such as inulin and lactulose can be added to food to improve nutrient absorption and health benefits such as calcium and magnesium (Choudhary & Tandon, 2009). Prebiotics have been shown to increase beneficial bacteria in the gastrointestinal tract, lowering immune-mediated food degradation.

A synbiotic is a probiotic and prebiotic combination found in foods, medications, and nutritional supplements for this article. Scientists discovered in 1960 that synbiotics could be used in food to change the microbiota in the human stomach. However, technological limitations hampered further development. Researchers are currently studying synbiotics and various metabolic disorders, such as diet-induced obesity, thanks to technological advances such as high-throughput sequencing. Probiotics and prebiotics are essential for improving human health, reducing diet-related illnesses, and providing specific nutrients. Dairy products must be consumed in addition to natural foods to meet the body's nutritional needs.

According to research, yoghurt consumption can help with lactose intolerance and constipation. Fermented milk containing *Lactobacillus gasseri* 505 and *Cudrania tricuspidata* leaf extract protected mice from liver damage. Probiotics and prebiotics in dairy products have been hampered in their use and development Because of a lack of research into the processes that lead to their therapeutic effects. More research is thus required in the field of microecological preparations, particularly in the areas of probiotics and prebiotics. This study looked at how ampicillin sodium stimulated two regulatory mechanisms that helped mice with gut microbiome imbalances restore their flora (Chalas et al., 2016). Allowing the body to heal itself was one method, while supplementing it with probiotics (*Lactobacillus rhamnosus* AS 1.2466) or prebiotics (xylooligosaccharide (XOS) and red ginseng extracts (RGE) was another. When we tested the four additions on animals with disrupted intestinal microbiotas, we discovered that they restored a variety of gastrointestinal tract (GIT) bacteria to varying degrees. Each new strategy was supplemented with GI tract microorganisms.

4.8 STIMULATION AND DEVELOPMENT OF THE IMMUNE SYSTEM

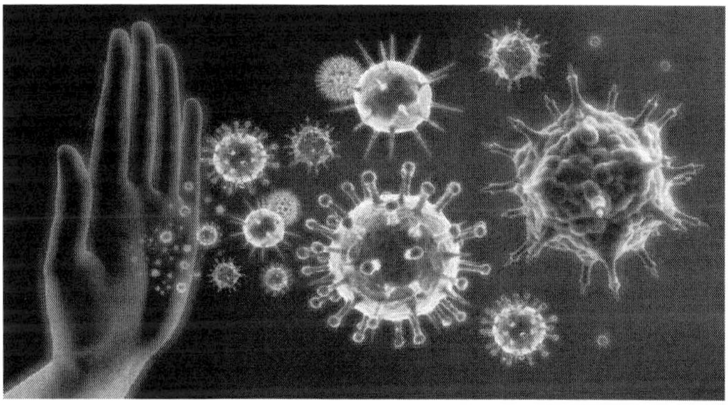

Figure 4.3: Illustration of immune system defense.

One of the many advantages of probiotic microorganisms is immune stimulation. This article goes into detail about how probiotics can help the immune system. Lactobacillus GG, *Bifidobacterium animalis* Bb-12, *Bifidobacterium lactis* DR10, *Lactobacillus johnsonii* La1, and *Saccharomyces cerevisiae boulardii* are widely reported to be the probiotic cultures with the greatest ability to alter immunity. The activation of macrophages, natural killer (NK) cells, antigen-specific cytotoxic T lymphocytes, and the release of various cytokines are all characteristics of the nonspecific cellular immune response that probiotics can boost in a strain- and dose-dependent manner. Different cytokine responses (Gram-positive and Gram-negative) may be produced depending on the content and type of probiotic bacterium (Gram-positive and Gram-negative). Probiotic supplementation during childhood may prevent the development of immune-mediated diseases, in contrast, probiotic intervention during pregnancy may affect fetal immunological parameters such as cord blood interferon (IFN), transforming growth factor (TGF), and breast milk immunoglobulin (Ig)A levels (Culligan *et al.*, 2009). Probiotics may improve gut mucosal immune function by increasing the number of IgA+ and cytokine-producing cells.

As one of the most microbiologically active ecosystems, the GIT contains a diverse range of microorganisms required for immune cell formation. Immune cells connected to the villi lamina propria coexist with microorganisms from the microbiome and those consumed in the diet. Although microbes do not interact directly with epithelial cells, their metabolites promote immune cell growth and activity.

Probiotics are bacteria that are beneficial to your health. They were classified as "live microbial feed supplements that aid the host's gut flora balance." As a result of this revision, "living bacteria" are now defined as "bacteria that have a positive effect on the host's health when administered in appropriate dosages." The intestinal microbiota contains a diverse range of probiotic bacteria, and an increasing number of these organisms are being added to foods to balance the gut microbiota and improve gut health. Probiotic microorganisms from the families Lactobacilli, Streptococci, and Pediococci are among the most commonly used. *Saccharum boulardii* yeast is also commonly used. Some bacteria strains, however, can be considered probiotic, but not all of them.

Animal studies were conducted to simulate the therapeutic effects of probiotics on host health and the treatment of a variety of viral and non-infectious illnesses. This supplement includes IBS treatment, *Helicobacter pylori* growth suppression, cancer prevention, gut inflammation reduction, and allergy protection (Cammarota *et al.,* 2014). More research is needed before probiotics can be effectively included in the human diet, health, and the management of various disorders such as diabetes, multidrug-resistant infections, and irritable bowel syndrome.

4.8.1 Probiotics Influence Immune System Stimulation Mechanisms

Probiotic bacteria must be able to tolerate the harsh environment of the human stomach and gastrointestinal tract to positively impact the health. The ability of probiotics to withstand bile salts and gastric juices may underpin claims that they are resistant to gastric juice, move through the upper digestive tract, proliferate and colonize, and function in the gut. Many probiotic bacteria were unable to function due to high levels of acidity in bile salts and stomach fluids (Charalampopoulos & Rastall, 2009). Probiotics can benefit human health by inhibiting the growth of potentially dangerous bacteria. Infections compete with probiotics for the resources they require to develop and reproduce. According to numerous studies, probiotics such as *Lactobacillus rhamnosus* strain GG and *Lactobacillus plantarum* can inhibit enteropathogenic *E. coli.*

To protect the host's connective tissue from the external environment, the intestinal epithelium undergoes numerous physical changes. This physical barrier is made up of epithelial cells, their close intercellular connections, and the mucus that covers the epithelial surface. The glycocalyx is another biochemical adaptation formed when Goblet cells secrete a mucin-rich and glycosylated layer. These chemicals combine to form an almost completely impermeable film on the epithelium's apical surface. According to research, taking probiotics orally increases goblet cells, which support the mucus layer. *Lactobacillus species* have also been shown in human intestinal cell lines to increase mucin production.

Probiotics benefit human health by inhibiting the growth of harmful bacteria, by producing bacteriocins with a high molecular weight and organic acids with a low molecular weight. Organic acids include lactic and acetic acids. These compounds have potent antibacterial properties against Gram-negative bacteria such as *Helicobacter pylori*. Bacteriocins are produced by bacteria such as *Lactobacillus acidophilus*, *Bifidobacterium bifidum* NCFB, *Lactobacillus plantarum*, and *Lactococcus lactis*. Paneth cells, which are distinct epithelial cells of the small intestine, produce antimicrobial peptides such as lysozyme, secretory phospholipase A2, defensin, defensin-like peptides, and cathelicidins. Synergy 1 and *B. longum* were discovered to increase defensin release from epithelial cells of patients with active UC. Pathogens cannot transfer from the bulky distal gut to the small intestine because of the intestine's unidirectional peristaltic motions (Bowe *et al.*, 2014).

Only a few studies have looked into enteroaggregative bactericidal response of *E. coli* to *E. faecium* supernatant, which is characterized by cell lysis and membrane damage. This bacterium produces enterocins, which are biopreservatives. Antimicrobial peptides, a new class of medicines, may be investigated in the future as a way to reduce resistance and protect the host through their specific antibacterial activity. Probiotics help to maintain intestinal microbial balance and prevent the growth of potentially harmful bacteria when taken as directed. Numerous studies have found that *Lactobacillus acidophilus or Lactobacillus casei* increased LAB while decreasing fecal coliforms and anaerobes. According to one study, consuming probiotics such as Prevotella and Oscillibacter altered the composition of the gut microbiota. As a result of these bacteria's anti-inflammatory chemicals, Th17 polarization is inhibited, and Treg/Type 1 regulatory T (Tr1) cells are formed in the gut.

Many different probiotic effects require the ability to withstand acid and bile excretion. Why are gastrointestinal epithelial cells more sensitive to probiotics although they are made up of microorganisms? Only when the probiotics are alive can the immune system be strengthened. IECs, macrophages, and dendritic cells attached to Peyer's patches or villi lamina propria interact with living bacteria and phagocytose probiotic cellular fragments attached to the PP or lamina propria of the villi; thus, bacteria that are no longer viable are quickly eliminated from the intestine (Ayichew *et al.*, 2017). Probiotics limit pathogen colonization through direct cell death, resource competition, and an enhanced gut-associated immune repertoire.

Oral probiotic therapy also protects the bronchial and urogenital mucosae from infection. The human papillomavirus, which causes cervical cancer, was eliminated more quickly after six months of frequent ingestion of probiotics. Oral probiotic treatment has been shown in animal models to improve macrophage phagocytic and microbic activity, lowering the risk of Salmonella infection. Lactobacilli probiotics help children and adults avoid antibiotic-induced diarrhoea.

The ability of the intestinal barrier to physically separate germs from the lumen, generate mucus, and immunoglobulin A, which play important roles in body protection (sIgA), is an essential feature (sIgA). The humoral adaptive immune system, particularly in mucosal areas, is heavily reliant on IgA antibodies. The majority of antibodies, which are primarily dimers linked together by a connecting chain, are produced by the intestinal lamina propria plasma cells. When IECs bind to the polymeric immunoglobulin receptor, dimerized IgA is transported into the intestinal lumen. Antigens in mucus are prevented from entering the body because of the ability of sIgA to adhere to the mucus layer. Not only does sIgA support local homeostasis in the gut lumen but also in the surrounding tissue. sIgA antibodies prevent toxins and pathogens from entering the digestive tract, a non-inflammatory process known as immune exclusion. Thanks to sIgA antibodies, DCs can more easily sample the subepithelial dome region of PPs. IgA antibody-secreting plasma cells in intestinal tissues are being studied to better understand their formation, distribution, and maintenance (Arora & Baldi, 2015). After probiotic delivery in this setting, IgA+ cells increased in the bronchi, breasts, and gut lamina propria. Probiotics have been shown to stimulate the immune system in other mucosal regions, where they can also increase and maintain immunological surveillance.

Fermented milk containing probiotics (PFM) has grown popular in recent years. By adding vitamins, proteins, essential amino acids, and essential fatty acids to dietary substrates, fermentation can improve the digestibility and nutritional value of food. The effects of probiotic cytokines on immune cells beyond the intestine (were studied using probiotic bacteria in PFM. PFM treatment increases the microbicidal and phagocytic activity of peritoneal and splenic macrophages. Probiotics work to boost the immune system and, consequently, antibody levels. Following a challenge with this pathogen, antibodies against *S. typhimurium* were critical in preventing the spread of dangerous germs to the liver and spleen. When the undernourishment hypothesis is considered, this effect becomes more pronounced.

Malnutrition is a condition that affects the entire body due to nutritional imbalance versus energy expenditure. As a result, the host's immune system is weakened, increasing the risk of infection. The effect of probiotics on the health of their hosts can be studied using a malnutrition model. As a re-nutrition meal, PFM restored the structure of the intestinal mucosa and improved local and systemic immunity in starved mice. The thymus, one of the body's defences against infection, is especially vulnerable to malnutrition's damaging effects. As a result, it is critical to use probiotics to restore thymus histology. The organ also showed a decrease in cell death and a recovery of single-positive CD4+ and CD8+ thymocytes. Several cytokines were elevated in the thymus of mice given PFM. There is some disagreement about the minimum effective dosage of probiotics, but it is typically accepted that at least 10^6 CFU/mL or grams of microorganisms should be consumed daily (Zhang *et al.*, 2016). Long-term PFM use has immunomodulatory effects that help maintain intestinal homeostasis without causing harm. The homeostasis of gut immune system was maintained and downregulated by cytokines like IL-10, which prevented an inflammatory immune response.

Probiotics are effective in treating allergic reactions. One of the most crucial indicators of this process is an increase in IgE. Probiotics lower immunoglobulin levels and reduce symptoms. However, the mechanisms underlying allergy relief remain unknown. The undeniable anticancer potency of probiotics is the result of multiple mechanisms working in concert. Probiotics have anti-genotoxic and anti-gene mutation properties, and the ability to reduce harmful bacteria in the gut. Probiotics affect the gut cells that control the immune system. Protecting the mucosal and stomach mucosal barrier and competing with pathogenic microorganisms, decreasing anti-inflammatory responses, degrading potentially carcinogenic chemicals, influencing cell proliferation and polyamine metabolism in the gastric mucosa, and other mechanisms

Numerous studies have shown that lactobacilli in dairy products boost the host's immune response. This trait has been discovered in *Bacillus longum*, *Lactobacillus casei subsp. rhamnosum*, and *Lactobacillus helveticus*. More research into the potential immune system benefits of probiotics is required. Consider lymphocyte proliferation, the production of interleukins 1, 2, and 6, TNF, prostaglandin E, serum total protein, albumin, globulin, and gamma interferon. Lactobacilli can naturally alter the immune system, making them useful in medical applications. Excess lymphocyte cytokine excretion and immunoglobulin-secreting cells in the blood are common signs of immune disorders. As part of an inflammatory immune response, activated monocytes and macrophages produce cytotoxic chemicals capable of lysing tumour cells *in vitro*. *S. thermophilus* 001158 and *L. bulgaricus* 100158 increased lymphocyte proliferation in the spleen, peripheral blood, and Peyer's patches of rats fed yoghurt containing both of these organisms, and IFN production in the spleen and patches (Yao *et al.*, 2020). In vitro, these cytokines kill and stabilize cancerous cells. Probiotics may be especially beneficial to the elderly because their immune systems weaken with age, making them more susceptible to disease. Regardless of the processes involved, probiotic cultures boost both non-specific and specific immunity. Probiotic microbes may be useful in treating, preventing infections, and cancer by boosting the immune system.

4.9 SYNTHESIZING AND ENHANCING THE BIOAVAILABILITY OF NUTRIENTS

4.9.1 Probiotics Increase the Bioavailability of Calcium

Figure 4.4: Foods High in Calcium.

Everyone is aware of how difficult it is to consume enough calcium consistently. Calcium supplements, contrary to popular perception, help preserve strong bones and lower the incidence of osteoporosis. It is critical to understand how much food is required each day to keep the bones strong. Osteoporosis and sarcopenia (loss of bone mass) are two disorders that hasten age (loss of bone mass). These elements have an impact on aging. If people want to live longer and better lives, they must treat sarcopenia and osteoporosis seriously.

A variety of factors influence calcium bioavailability in humans. The most popular classifications are exogenous and endogenous. Exogenous factors affect the body while, food has an external effect. The amount of calcium the body can absorb from the digestive tract is determined by the mineral's properties, such as its type and quantity, oxidation status, solubility, the presence or absence of antagonistic (competitive) ions, and the presence or absence of chemicals that facilitate or inhibit mineral absorption. Individual variances are influenced by genetic characteristics, age, gender, physiological conditions (such as pregnancy or nursing), and the amount and quality of mineral deposits in the body (Wescombe *et al.*, 2012).

Probiotics boost bone health by enhancing the body's calcium absorption capacity. Continue reading to find out more about how probiotics affect bone health. Probiotics slow bone loss by dissolving calcium and decreasing parathyroid hormone levels (greater PTH levels increase bone resorption by stimulating osteoclast activity). Wheat may contain phytate, which may reduce the quantity of calcium available. Probiotic bacteria produce the enzyme phytase, which permits actively blocked calcium to be released, boosting calcium availability at the absorption site. This is because Lactobacillus and Bifidobacteria in the gut break down the glycoside linkages in estrogenic meals, increasing calcium bioavailability. Calcium absorption has been linked to the vitamins folate, C, D, and K. Probiotic bacteria impact regulation of gut metabolic activity and vitamin production. Vitamins are commonly recognized as playing a role in this process.

Depending on the probiotics used, they can produce bioactive peptides. *Lactobacillus helveticus* produces bioactive peptides termed IPP and VPP, which contain proline and may help the body absorb calcium. Minerals may be liberated from inorganic ions that are not broken down by difficult-to-digest peptides. Ang II synthesis can be inhibited by valyl-prolyl-proline (VPP) and isoleucyl-prolyl-proline (IPP) (Ang I). Angiotensin II, a bone-resorbing hormone, has been associated with bone vasoconstriction. The fermented broccoli-derived probiotic *Bifidobacterium longum* has been demonstrated to reduce NF-kB gene expression, oxidative stress, and bone loss. According to one study, oxidative stress promotes osteoclast formation and bone resorption. In clinical trials, probiotics lowered intestinal inflammation while increasing bone mineral density (Valero-Cases *et al.*, 2020).

The lowering of TNF and IL-1 levels in the jejunum and ileum of the intestines, as well as using a probiotic strain (BMD), increased bone mineral density (BMD). The microbiota in the stomach has a significant impact on bone health. Because probiotics are required for cell proliferation, a recent study discovered that when ingested in appropriate quantities, they can significantly impact bone health. Calcium absorption, for example, may be hampered if the gut mucosa is injured, resulting in severe GI tract problems. Probiotics can help bone growth by generating metabolites, enzymes, and vitamins required for calcium metabolism and bone matrix formation. As a result, sickness and bone loss are on the decline. Numerous studies have demonstrated that probiotics boost bone health.

The effects of two well-known probiotics, *Lactobacillus sali* and *Bifidobacterium infantis*, on calcium absorption and transepithelial calcium transport, were studied in human intestinal-like Caco-2 cells. Calcium transport in UCC 118 and *E. coli* was significantly increased compared to the control. On the other hand, bacteria do not promote intestinal calcium transfer; instead, they promote absorption. Calcium, magnesium, phosphorus, and zinc were determined in fermented goat and cow milk products (Tadesse, 2012). This product was also tested for the probiotic bacterium *Lactobacillus fermentum* D3. Fermented milk containing probiotic bacterial strains demonstrated much higher calcium, phosphorus, and zinc bioavailability than commercially available fermented goat and cow milk products.

According to *in-vitro* research, strain D3 improves mineral bioavailability over other commercially available milk-fermented products. Mineral complexation is linked to phytases, an enzyme that breaks down phytate. Phytic acid's chelating effect on mineral absorption can be reduced by employing enzymes that dephosphorylate phytic acid. According to the findings, LactobacillusLactobacillus casei was genetically changed utilizing probiotics such as bifidobacteria. Phytase from *Bifidobacterium pseudocatenulatum* was produced, exported, and linked to the cell wall of *L. casei* utilizing a nisin-inducible promoter. According to the study, enzymes generated by bifidobacteria and *Lactobacillus casei* can reduce phytate levels in fermented foods. According to multiple studies, osteoporosis and sarcopenia, two of the most common aging-related illnesses, can be addressed with a calcium-rich diet (Smolyansky, 2010). Clinical trials have shown that patients with osteoporosis and osteopenia respond significantly to treatment after only six months.

Probiotics are supplements that contain beneficial bacteria. The bulk of probiotics in yogurt, fermented milk, and other fermented foods are produced by lactic acid-producing bacteria. Lactic acid bacteria-rich foods have various health benefits, including improved nutrient absorption, treatment of lactose intolerance symptoms, reduced allergy prevalence in allergy-prone individuals, and a lower risk of cancer. When probiotic bacteria are present, they create antimicrobial compounds that resist pathogens, compete for resources and growth factors with pathogen binding sites, activate immune modifying cells and make lactase. Probiotics can perform a variety of functions, including these. The list of benefits connected with probiotic usage is endless. This study looks at the efficacy, sources, and safety of probiotics. According to current studies, probiotics are necessary for both people and animals. Antimicrobial probiotics may be an effective strategy to avoid microbial disease because they are risk-free, inexpensive, and "natural." According to the study, experts are looking into whether probiotics can help with health maintenance and disease prevention.

Compared to rest, skeletal muscle energy turnover can be raised by 20 to 100 times during physical activity, training, or performance. A higher caloric and macronutrient intake is required to compensate for energy deprivation, replenish glycogen stores, stimulate muscle growth, and heal tissues. Athletes competing in sports such as weightlifting, jogging, and endurance competitions have nutritional restrictions and macronutrient requirements (Stavropoulou & Bezirtzoglou, 2020).

The human body requires proper amounts of vitamins, minerals, and trace elements to maintain optimal health, all of which must be present in the diet. If ingested in adequate quantities daily, micronutrients may not affect an individual's ability to exercise. Research has proven that young athletes require fewer micronutrients than non-athletes. In studies, comparing the micronutrient requirements of athletes versus non-athletes generated mixed results. Excessive perspiration is caused by exercise and athletic training. Sport intensity, gender, and individual variability are only a few factors that influence sodium (Na), the main component of sweat. Sweat-induced deficits in vitamins and trace elements appear to be minor, even during prolonged periods of considerable evaporation.

A healthy, balanced, and junk-free diet can help to prevent widespread vitamin shortages. Daily micronutrient requirements are not met immediately if more calories are consumed. The gut flora is dependent on bioavailability, food processing, and absorption. According to the research, bacteria improve mineral absorption, which benefits both hosts and animal and human gut microbiomes. Probiotics improve mineral absorption, which benefits the stomach microbes. Soaked and germinated vegetables and fermented cereals, are likely to include natural mineral additions that benefit human gut flora and overall health (Saraf et al., 2010). When ingested as "nutrition supplements," they can boost the solubility and bioavailability of micronutrients. More research is needed to understand how gut bacteria affect mineral bioavailability in humans and the mechanisms involved.

4.10 REDUCING SYMPTOMS OF LACTOSE INTOLERANCE

Lactose cannot be digested by more than 60% of the human population due to a lack of lactase enzyme activity. Probiotics are live bacteria or yeast that are introduced to the intestinal flora. Studies show that probiotics provide various health benefits, including better gut health, immune activation, and blood cholesterol reduction. Probiotic bacteria in fermented and unfermented milk products may help reduce lactose intolerance symptoms (LI). This systematic review examined 15 randomized, double-blind studies to assess whether probiotics are useful in treating diarrheal disease (LI). The eight most scientifically confirmed probiotic strains were evaluated (Sistek et al., 2006). Probiotics appear to have a substantial association with lactose intolerance, with varying degrees of efficiency.

4.11 WHAT EXACTLY IS LACTOSE INTOLERANCE?

Figure 4.5: Lactose-free milk.

Lactose, a natural sugar found in milk and dairy products, cannot be digested by the body. Lactose is transformed in the small intestine by the lactase enzyme into glucose and galactose, two monosaccharides easily absorbed by the body. When there is insufficient lactase in the small intestines, lactose is broken down by bacteria in the colon (big intestine). Short-chain fatty acids and gases such as h2, co2, and h3 are produced during this metabolic pathway. Lactose intolerant people frequently experience digestive problems such as gas, bloating, and cramping. Lactose that has not been digested might cause increased stomach acidity and osmotic pressure (Sidhu *et al.,* 2015). The duration of lactose intolerance symptoms varies from person to person and is determined by the amount of lactose taken and the time it takes for the stomach to empty.

4.11.1 Probiotics May Aid Lactose Intolerant People

According to various studies, bacteria present in yogurt and nondairy probiotic supplements may aid with lactose digestion. According to various studies, fermented dairy products with live cultures, such as yogurt, are more pleasant to lactose-intolerant people than milk. Lactose is more extensively broken down in the small intestine because fermented foods impede stomach emptying, allowing lactose to enter the colon more slowly. The probiotic bacteria-produced enzyme beta-galactosidase aids lactose digestion in the colon. These two components may make fermented dairy products simpler to digest. Lactose fermented with active microorganisms is easier for the digestive system to digest.

Lactose intolerant individuals should consume probiotics and fermented foods. Is there a specific type of probiotic that can help people who have lactose intolerance?

4.12 THE BEST LACTOSE-INTOLERANT PROBIOTICS

Probiotic supplements were given to lactose-intolerant patients to see how their gut flora responded. Lactose intolerance and colon metabolic activity improved after only two weeks of yoghurt (which contains higher *Bifidobacterium animalis*) and *Bifidobacterium longum* supplementation. Furthermore, the two-week therapy boosted fecal-galactosidase activity. Lactose intolerance can be relieved by taking probiotic tablets or eating foods containing live cultures, which can help balance the microbiota and reduce lactose intolerance and digestion.

Patients' intestinal regularity and dairy sensitivity were enhanced when *Lactobacillus rhamnosus* Rosell-11 and *Lactobacillus acidophilus* Rosell-52, two Lactobacillus strains, were introduced. In a double-blind, placebo-controlled study, Lactobacillus acidophilus DDS-1 reduced lactose intolerance symptoms such as diarrhea, vomiting, gas, and stomach cramps. Each of the 38 lactose-intolerant participants received a placebo and a probiotic supplement containing 10 billion CFUs of *Lactobacillus acidophilus* DDS-1 as part of this two-arm crossover research (Singh et al., 2011). There was a statistically significant improvement in diarrhea, stomach cramps, vomiting, and overall symptom scores compared to the placebo group. The supplementation of *Lactobacillus acidophilus* LA-05, *Lactobacillus acidophilus* NCFM, and *Bifidobacterium lactis* BB-12 has been shown in several studies to enhance symptoms and dairy product tolerance.

CHAPTER 5

PROSPECTIVE APPLICATIONS OF PROBIOTICS

CONTENTS

To benefit the health of their hosts, probiotics are live bacteria added to food. Included in dairy products for centuries are these ingredients. As technology has advanced, it has become possible to examine its applications in greater depth. The suppression of pathogens by bacitracin production is a widely reported characteristic of antibacterial activity. Research suggests that probiotics may be a better alternative to antimicrobial drugs because antibiotics may promote resistance in bacteria (Singh & Natraj, 2021). Studies show that probiotics reduce lactose in dairy products, improve digestion, and aid in animal growth and development. Lower cholesterol and an increase in cytokines (and IgA) have also been linked to probiotic consumption.

Various *in vitro* and *in vivo* studies on the effects of probiotics on the human body, as well as current technological applications and potential future probiotic developments.

After deconstructing and evaluating the key strains that may be included in probiotic capsules, it is simple to grasp why they are effective for safeguarding the general wellness._They give various other health advantages. Prebiotics are key components of probiotic supplements that support and promote health in practically every body system.

1. A balanced mix of "good" and "bad" bacteria in your stomach aids optimal digestion and nutrient absorption.
2. Feeding "good" bacteria that may occasionally be overwhelmed by oxidative stress.
3. Promoting proper peristalsis, which turns food into nutrients that the body can use for energy, cell development, and repair.
4. Promoting mineral absorption from the meals, particularly those that support bone health, such as iron, calcium, magnesium, and zinc.

Consumer demand for probiotic foods has increased due to increased public awareness of diet-related health issues and increasing evidence of the health benefits of probiotics. Probiotics can be found in yogurt, frozen dairy desserts, spray-dried milk powder, cheeses, ice cream, freeze-dried yogurt, and fruit juices. It has been proposed that 109 CFU of probiotic microorganisms be consumed daily to reap probiotics' benefits (Rusu *et al.*, 2019). A product should have at least 107 probiotic cells per gram or mL of food, which is also recommended in Japan, based on daily consumption of 100 g or mL of probiotic food.

Probiotics have traditionally been found in dairy products such as milk and yogurt. More than one study has found a deficiency of probiotic bacteria cells in commercial yogurt. Viable probiotics can be delivered to the gastrointestinal tract through cheeses rather than fresh fermented products (such as yogurt). Probiotic bacteria may be protected during transit through the GI tract if the cheese has a higher pH and a more solid consistency. Cheese has a higher buffering capacity than yogurt. When adding probiotics to foods, selecting a compatible probiotic strain and food type combination is critical. Other important considerations include using food processing conditions that support probiotic survival, ensuring the food matrix supports probiotic growth (if fermentation is required) and selecting a product matrix, packaging, and environmental conditions that ensure adequate probiotic survival throughout the product's shelf life.

Specific probiotic species may:

- help with digestion
- stop diarrhea
- cure genital infections
- stop autoimmune diseases
- improve skin problems
- prevent urinary tract infections

It is crucial to bear in mind, though, that not everyone will benefit from every probiotic in the same manner. Some physicians recommend mixing probiotic pills with antibiotics to address a specific condition. As a result, both healthy and hazardous germs could be eliminated, which might create diarrhea. Probiotics help in maintaining or restoring a healthy microbiome in the gastrointestinal system.

5.1 CONFECTIONARY

Food items that give extra health benefits in addition to nutrition are becoming increasingly popular as people become more aware of the link between a balanced diet, lifestyle, and overall health. Maintaining human health and wellness is dependent on the consumption of the suitable probiotic strains, according to research. A probiotic is a probiotic microorganism (Abuajah *et al.,* 2015). Probiotics, as defined by the FAO/WHO, are "living microorganisms that give a health benefit to the host when fed in suitable amounts." Probiotics have sound effects on the body, including the maintenance or stimulation of several physiological processesand the decrease of symptoms and risk factors associated with certain health issues.

Probiotics have been linked to many health benefits, including improved gastrointestinal health and modulation of the immune system. New insights into the human body's health are emerging from studies on the impact of probiotics. Probiotics are rapidly being included in unique functional food products as more knowledge becomes available. This may be the case with the rise in popularity of probiotic-enriched sweets. The critical components for microbial survival in confectionery applications must be understood to create efficient probiotic-containing sweets.

5.2 EXTRUDED SNACK

Researchers have been studying microbial cultures for their ability to prevent and treat various diseases since the turn of the century. Their use in fermentation of food and drink dates back to ancient times. Thus, the terms "pro-life" and "probiotics" were coined. Clinical experiments on the efficacy of probiotics on constipation began in the 1930s, and research has progressed ever since. Probiotic bacteria can shift the balance of healthy and bad microbes in the digestive system by inhibiting potentially hazardous pathogens. There are now many foods and pills that include probiotics. Bacteria such as Lactobacillus and Bifidobacterium are frequently found in these foods. Commercially available probiotics include a wide range of different species (Aggarwal *et al.,* 2013). Dietary supplements that include microorganisms designed to improve the health and well-being of the customer are called "probiotics." Traditional yogurts, bio-yogurts containing unique bacteria from the intestines, and other health-enhancing products make up this collection. An even more precise description would be "a live microbial dietary supplement that boosts the host animal by regulating its gut flora,"

Temperature, humidity, high pH levels, organic acids, and other physical and chemical variables can have a detrimental effect on probiotic microorganisms in food and feed products. Heat treatment is used in many food and feed manufacturing processes, reducing the viability of probiotic bacteria both during production and after that. To ensure the viability and stability of probiotic bacteria, food and feed must be processed via these processes first. As a result, probiotic-enriched food and feed products should not be prepared after the inclusion stage. In addition, product compositions should not harm the viability of probiotics by including chemicals. Probiotic bacteria may now be added to extrude dry meals due to recent technological advancements (Reid, 2016).

In the food industry, extrusion is a common practice. Shapes can be created by moving material through a region of high pressure and temperature while increasing shearing. Extruded materials can use these process parameters if desired, and a wide range of end products can be made by varying combinations. A wide variety of materials may be extruded, allowing for a wide variety of textures and shapes. Extrusion was used to make pasta, snacks, pet food, and even ready-to-eat munchies. Snack and breakfast food manufacturers can benefit from twin-screw extruders because of their excellent mixing capabilities, high screw speeds, and reasonable throughput rates (Shah and Gupta, 2004). Extrusion has been used to make a variety of snack meals, including rice, rice, and black bean cookies, pellets consisting of wheat, rice, and potatoes, foxtail millet-based snacks, chickpea, and rice flour blended snacks, apple, beetroot, carrot, cranberry, and gluten-free teff flour cereal. Today's food and feed manufacturers and markets are still unsure how to handle extruded foods and feeds containing probiotics. As a result, it is critical to look at how probiotics might improve the value of extruded goods in the marketplace (Qi *et al.*, 2019). Some experts believe non-proliferating probiotic bacteria can be included in the extrusion process. Biologically inert, non-viable, non-replicating, and/or present in DNA, metabolites, cytoplasmic chemicals, or cell wall components are all examples of non-replicating probiotic microorganisms.

Because of the rising number of manufacturers using technological advances to develop new goods, the market for extruded snacks has grown dramatically in the last few years. Several new items have entered the market to join the ranks of well-known category mainstays thanks to the novel barrel and die designs for twin-screw extrusion. Co-extruded snacks are becoming increasingly popular among consumers and snack manufacturers. Breakfast cereals are a subcategory of this (products that often are consumed as snacks). This technology has been around for a while, but Shaw points out that it has recently grown in popularity. The conventional distinctions between snack, cookie, cracker, and cereal are blurred with this product.

Consumer preferences may be tested and analyzed due to the wide range of snacks on the market. With the flexibility of twin-screw extrusion technology, companies may experiment with different extruder settings to create meals with a wide range of nutritive properties, according to Shaw.

The use of a twin-screw extruder provides this essential benefit. Changing the screw arrangement, screw speed, liquid and/or steam injection, and die design can alter the amount of thermal and/or mechanical cooking (Aponte *et al.,* 2020). Modern extruders can generate a wide variety of foods for people of all ages and interests, from colorful co-extruded meals to intricate dies for expanded and unexpanded goods. An ever-expanding range of applications may be achieved through extrusion technology.

"Half-products," as they have come to be referred to, have proliferated because of this (multidimensional products subsequently puffed, popped, or fried). Second-generation products and products for immediate expansion Delivering new and exciting eating experiences has become an increasingly important goal for 3G-type items. The texture and flavor of a 3G snack may vary depending on how it is puffed, even if the recipe is the same (hot air, microwave, oil, etc.). The development of extruded snacks has dramatically benefited from in-depth ingredient research and development. Twin-screw technology is beginning to be recognized by snack manufacturers as a potential advantage since it allows them to produce snacks with a range of components using existing technologies. New drugs have been created in response to the high need for food. Extruded snacks have grown "healthier" by including raw foods such as cereals, ancient grains, legumes, and some vegetables.

Baked rather than deep-fried meals are healthier because of extrusion's ability to bake rather than deep-fry. A wide range of cereals, including whole grains, work well with extruded snacks. Creating snacks that are both healthy and useful necessitates the use of intelligent components.

5.3 SWEETENER

There is a lot of sugarcane Saccharum in the world. An increase in rootstock growth. The stalk's height ranges from three to six meters. People's Republic of China Guangxi, Sichuan, and Yunnan are just a few of the many desert provinces in southern China where it is customary to produce Sugarcane that thrives in the sun and well-drained soil. The severe temperature differential between the summer and winter months at the feet. Ethanol may be made from sugarcane, a tropical and temperate plant. It can also be used as a raw material to make sucrose (Prins, 2011). Sugarcane is grown in over a hundred nations, with Brazil, India, and China supplying most of the world's supply. It is scrumptious. Sugarcane is rich in water, vitamins, fat, and protein, all essential to human metabolism. Sugaring is made up of organic acids, calcium, and iron, and the epidermis may be seen in various hues, including purple and green.

5.4 PROBIOTICS (PREBIOTICS)

Bacterial colonies are encouraged to develop by taking this supplement. Increasing the amount of food that cannot be digested helps the host's health. Thus, microbial growth, activity, and a positive effect are encouraged. Dietary fiber accounts for the majority of its composition. Most of the gut flora will not be digested by the upper digestive system; therefore, successful probiotics will ferment the remaining fraction. It is essential to encourage the development of profitable strains rather than corrupt or destructive ones. Bacteria that thrive in a stressful environment

The composition of functional meals exposes how the body hinders people from exercising successfully, which is why they may help improve health care. Illness prevention and rehabilitation are only a few of the many functions of the imperial function. It is possible to address energy-related studies and the creation of functional foods. Increased dietary intake, prevention and treatment of non-communicable chronic diseases and an overall improvement in nutritional status are unquestionably essential needs in China. Physical characteristics of the functional inhabitant foods rely on oligosaccharides, which are difficult for human digestive enzymes and stomach acid to break down but are also challenging for the intestines to absorb (Pratap *et al.*, 2020). They can breed bifid bacteria and lower blood pressure and fat levels without altering physiological processes like blood glucose. Now, more local and worldwide oligosaccharide research is being conducted.

The principal oligosaccharides are oligoisomaltose, xylooligosaccharides, and soy oligosaccharides. Carbohydrates include sugar and fructooligosaccharides like galactooligosaccharides. However, the amount of oligosaccharides generated in China and other affluent countries is vastly different. Research on oligosaccharides is progressing rapidly, it is widely used in food processing, and it typically lives up to customer expectations. A healthy and well-balanced diet is essential for good health. According to the current invention, adding probiotics to a sweetener can improve its nutritional and health advantages. When it is used in the usual way of sugar production, the prebiotic bacterial content of its sweetener is limited.

5.5 COMPOSITION OF THE NEW IDEA

Natural bacterial sweeteners may significantly improve the flavor of sweet foods, and this idea intends to give a method for making them. The prebiotic bacterial component of the agent increases the sweetener's form point and nutritional value by converting sugarcane's soluble dietary fiber to insoluble daily fiber.

A probiotic sweetener can be made using the following methods:

Crushed cane, cellulase, and phosphoric acid are all mixed. Many processing methods are employed to create Gao Yi Sweetness, including "beta"-mannose secondary enzymolysis, *Trichoderma viride* secondary fermentation, Zygosaccharides fermentation sterilization, and filtering (Pandey *et al.,* 2015).

In addition, the probiotic sweetener production process includes the following processes:

Sugarcane juice and bagasse are extracted from the fresh cane using a press.

A phosphoric acid and cellulase solution is used to break down the cellulose in bagasse and sugarcane juice. After twenty-six to eight hours of digestion at a temperature of forty °C, the sugarcane mixed liquid of primary enzymolysis is formed.

Zygosaccharomyces fermentation will be used in the fermentation. Initial enzymatic hydrolysis sugarcane combined liquor is put into yeast starting liquid and left to ferment for one to three hours. Zygosaccharomyces fermentation mixtures are then made from sugarcane.

Sterilization was performed on the sugarcane-infused beverage. It takes 15 to 30 minutes to digest beta '-mannose in a sterilized sugarcane solution using the "beta '-mannose secondary enzymolysis" method. Secondary enzymolysis uses activated *Trichoderma viride*, which is then added to a sugarcane-mixed liquid and fermented for 28 to 50 hours at a temperature of 26 to 3 °C.

A press device removes juice from the secondary enzymatically-mixed sugarcane liquid, which is then filtrated to reach high prebiotic concentrations. Pathogen-infected sugarcane juice Sugarcane liquor with a high concentration of probiotics is cooked for five to six hours to extract the probiotics. Then water is added to the mixture (Parvez *et al.*, 2006).

Inject cellulase from 0.01 to 0.8 % (m/V) into the solution. Phosphoric acid is also included, ranging from 0.02 to 0.15 % (m/V). Zygosaccharomyces fermentative bacteria are present in the inoculum in concentrations ranging from 0.3 % to 1.2 %.

There is also between 0.4 % and 0.5 % beta-mannanase included. Additionally, the *Trichoderma viride* inoculum concentration ranged from 2 to 10 %.

5.5.1 Innovative Benefits at This Time

Prebiotic bacteria are present in the sweetener produced by this invention's method, which includes cellulase, phosphoric acid, Zygosaccharomyces fermentation, "beta"-mannose secondary fermentation, and *Trichoderma viride* secondary fermentation. Sweeteners' conventional properties and advantages, as well as their prebiotic bacterial content, might be greatly enhanced by the current innovation. There is a 0.74 % decrease in the quantity of fiber ingested. For those who consume other harmful and weakening meals, the raw bacterium sweetener enhances the sweetener's nutritional content significantly. It has an overall beneficial effect on the blood, especially for those who drink other poisonous and weakening foods (Novik & Savich, 2020). Extracted sugarcane juice is blended with sugarcane juice and passed through the fiber by the present invention.

The quantity of nutritional fiber in the final sweetener can be increased by simple enzyme and phosphoric acid processing. After handling Zygosaccharomyces fermentation, "beta"-mannanase secondary enzymolysis, and *Trichoderma viride* secondary fermentation, the procedure is finished. The amount of soluble dietary fiber was discovered for the first time experimentally using *Trichoderma viride*. The soluble dietary fiber content of the fermentation sweetener is 3.59.0.28 %, which is much lower than the inventions.

Sugarcane's nutritional component is maximized to maximize the creative approach's raw components, manufactured in a pure day-right plant and conform to national restrictions on food additives. Taste compounds have an enormous economic potential because they boost sweeteners' nutritional value.

5.5.2 Embodiment

It is important to note that while certain embodiments of the present invention will be discussed in further depth, these embodiments serve as examples of this method. The scope of the current invention must be broadened rather than restricted, as individuals who are competent in the area would identify other uses for it upon reading it (Nichols, 2007). The application's claims include a change in valency form.

For example, China General Microbiological Culture Preservation Administrative Center, CGMCC number: 3.2942, provides the *Trichoderma viride* reference culture and Zygosaccharomyces fermentation bacterial strains for purchase, respectively, in one embodiment of the present invention.

Figure 5.1: Fermented food collection.

In addition, substances approved for human consumption are used. When God first took on human form

Methods and instruments for making probiotic sweets are included in the current embodiments. The body is made up of the acts listed below:

Squeezing sugarcane juice from fresh cane produces bagasse and cane juice.

Cellulase and phosphoric acid are then added:

In this case, sugarcane juice is mixed with a 0.2 % cellulose enzyme and 0.1 % bagasse concentration (Narang *et al.*, 2011).

Phosphoric acid is digested at 30 °C for 7.5 hours to extract the primary enzymatic liquid from sugarcane.

Zygosaccharomyces fermentation is responsible for the fermentation of the final component, carbon dioxide.

Zygosaccharomyces fermentation is introduced into a PDA liquid medium activation culture from sugarcane mixed liquor infected with Zygosaccharomyces fermentation. This results in a Zygosaccharomyces fermentation seed fluid spore concentration of 5.0 106 Individual/mL.

A Zygosaccharomyces fermentation-fermented sugarcane mixture is sterile after 25 minutes at 65 °C high-temperature processing.

To make secondary enzymatically digested sugarcane mixed liquid, 0.46 % of the sugarcane mixed liquor (m/V) is added to the m/V. After 20 minutes of digestion at 30 °C, the sugarcane mixed liquid is ready (Nami *et al.*, 2015).

Subsequently, the sugarcane-mixed liquid is used to inject the activated *Trichoderma viride* into the secondary enzymolysis fermentation. Fermentation at 26 °C and a 5 % inoculum concentration require 40 hours.

A filter medium with 0.3 mm diameter holes is used to remove insoluble particles from the sugarcane mixed liquid from secondary enzymolysis. So, sugarcane is utilized to make alcohol.

The sugarcane blended liquor with a high concentration of probiotics is boiled in the seventh stage.

Conventional sweetener production procedures are used to combine sugarcane with high microbiological content. Swirling continually while boiling for 5 to 6 hours, the high probiotic sugarcane blended liquor loses its moisture. The probiotic sweetener of the present invention is made by gradually increasing the sugar content in the combination (McKenzie *et al.*, 2018).

The inventor is asked to confirm that the prebiotic bacterial content of a sweetener is produced using an enzyme-gravimetric approach. GB/T5009.88 Determine the amount of dietary fiber in the sweetener.

In contrast to probiotic sweeteners, which comprise 18.45 0.74 % soluble fiber, standard sweeteners have less than one per cent soluble fiber.

The novel approach may greatly boost the number of probiotics that can be added to the sweetener.

The second illustration

Methods and instruments for making probiotic sweets are included in the current embodiments. The body is made up of the acts listed below:

Bagasse and cane juice is produced when sugarcane juice is squeezed from fresh cane.

Then cellulase and phosphoric acid are added:

Combine bagasse and sugarcane juice with 0.01 % (m/V) Fiber Plain enzyme and 0.02 % (m/V) Fiber Plain enzyme (Arora & Baldi, 2015).

Phosphoric acid is digested at 20 °C for six hours to extract the primary enzymatic liquid from sugarcane.

Zygosaccharomyces fermentation is responsible for the final component, carbon dioxide, fermentation.

To produce, Zygosaccharides fermentation is introduced into an activation culture and PDA liquid medium. The inoculum contains 0.3 % Zygosaccharomyces fermentation, and the sugarcane-mixed fluid of Zygosaccharomyces fermentation is removed after one hour. Changing the Zygosaccharomyces fermentation spore concentration in Zygosaccharomyces fermentation seed liquid Individual/mL takes 2.0 106 Zygosaccharomyces fermentation spores per milliliter.

The sugarcane mixed liquor fermented by Zygosaccharomyces fermentation is placed in a 60 °C environment and sterilized after 10 minutes of high-temperature processing.

Fifth, 'beta '-mannose is digested: 0.4 % of the sugarcane mixed liquor of sterilization (m/V)'beta '-mannose is added to the sugarcane mixed liquor of secondary enzymolysis, and 15 minutes at 20 °C are digested to generate the sugarcane mixed liquor of secondary enzymolysis.

Sixth, secondary fermentation of *Trichoderma viride*: After activation, *Trichoderma viride* is added to the sugarcane mixed liquid for secondary enzymolysis. Fermentation lasts 28 hours at a temperature of 26 °C with a 2% *Trichoderma viride* inoculum (Mahasneh & Abbas, 2010).

The sugarcane secondary enzymolysis liquid mixture is placed in a press machine, and any insoluble material is removed using a filtering medium with a pore size of 0.5 mm. As a result, sugarcane is used to produce alcohol.

In the seventh stage, the sugarcane blended liquor with a high concentration of probiotics is boiled.

Sugarcane with a high microbiological content is combined using conventional sweetener production procedures. The high probiotic sugarcane blended liquor loses moisture by constantly swirling while boiling for 5 to 6 hours. The present invention's probiotic sweetener is created by gradually increasing the sugar content of the combination.

Third demonstration

The current embodiments include methods and instruments for making probiotic sweets. The body is made up of the following acts:

Bagasse and cane juice is produced when sugarcane juice is squeezed from fresh cane.

Then cellulase and phosphoric acid are added:

Primary enzymolysis sugarcane mixed liquor is made by combining bagasse and sugarcane juice, adding 0.8 % (m/V) cellulose enzyme and 0.15 % (m/V) phosphoric acid and digesting the mixture for eight hours at 40 °C (Li *et al.*, 2016).

Third, Zygosaccharomyces fermentation ferments: the bacterium produces alcohol after being inoculated and placed in an activation culture of a PDA liquid medium. Individual/mL Zygosaccharomyces fermentation spore concentration in Zygosaccharomyces fermentation seed fluid must be increased to 8.0 106. The Zygosaccharomyces fermentation seed liquor must be seeded in a sugarcane mixed liquor with a 1.2 % inoculum concentration.

After 30 minutes of processing at a high temperature, the sugarcane mixed liquor fermented by Zygosaccharomyces fermentation is heated to 80 °C and sterilized.

Fifth, beta-mannanase is digested: 0.5 % sterilizing beta-mannanase (m/V) is added to the sugarcane mixed liquor, which is then digested at 40 °C for 30 minutes to produce secondary enzymolysis sugarcane mixed liquor.

The activated *Trichoderma viride* is injected into the secondary enzymolysis fermentation using the sugarcane-mixed liquid. Fermentation takes 50 hours at a temperature of 26°C and an inoculum concentration of 10%.

The juice's pore diameters and filtering medium are removed after the sugarcane secondary enzymolysis liquid combination is placed in press equipment. To obtain sugarcane combined liquid, insoluble components with particle sizes greater than 0.3 mm must be filtered through 0.1 mm.

In the seventh stage, the sugarcane blended liquor with a high concentration of probiotics is boiled.

Sugarcane with a high microbiological content is combined using conventional sweetener production procedures. The high probiotic sugarcane blended liquor loses moisture by constantly swirling while boiling for 5 to 6 hours (Lin *et al.,* 2019). The present invention's probiotic sweetener is created by gradually increasing the sugar content of the combination.

5.6 DAIRY, INFANT FORMULA, AND BEVERAGE

Milk and its derivative food matrices are critical for constructing healthful meals. Now that their probiotic characteristics have been revealed, fresh isolates of probiotic-capable autochthonous lactic acid bacteria are more likely to be discovered. Cheeses, fermented dairy products, ice cream, and baby formula have all been thoroughly researched to create unique functional meals. The ability of probiotic foods to restore the gut microbiota, which is vital for maintaining the balance between health and disease processes, benefits both firms and customers (Kumar, 2013). The discovery and development of recombinant microorganisms with antiviral properties have advanced in recent years. Looking into probiotics in the context of milk was a key element of this chapter's focus. Microorganisms, products, and strategies for developing innovative, risk-free functional foods are also covered.

The phrase "probiotics" may be discussed when shopping for infant formula and supplements (which means "for life"). Probiotics are specific types of live bacteria that may be used as an ingredient in some recipes. Breastfeeding women might benefit from probiotic drops or powders as prescribed by their doctors. The digestive tracts of breastfed neonates are teeming with "friendly" microbes.

One of the key advantages of probiotics is that they prevent the multiplication of "hostile" microbes in the intestines of formula-fed babies.

In recent years, there has been an upsurge in the availability of probiotic supplements, particularly for breastfeeding neonates. Some doctors prescribe probiotics for children born through C-sections or whose mothers were given antibiotics during labor, and researchers are presently studying the advantages of this medication (Kukkupuni *et al.,* 2015).

The two most frequent probiotic strains in the world are lactobacilli and bifidobacteria. Probiotics may help cure viral diarrhea in children and pediatric atopic dermatitis, according to research (eczema). An ongoing study is looking at health advantages such as a lower incidence of food allergies and asthma, avoiding or alleviating urethral infections, and decreasing newborn colic symptoms.

We need a lot more study before we can safely declare that probiotics can assist with any health issues. When taking probiotics, it appears that the advantages will last as long as feasible. When you stop administering probiotic-enriched formula to your infant, the flora in his or her digestive system returns to normal. On the other hand, breastfed babies have more adaptable gut flora, which results in better health outcomes.

In addition to their intrinsic qualities, milk and milk derivatives are effective probiotic delivery carriers since they are frequently stored in freezers. Probiotics may be present in many commercial dairy products, including fresh and sour milk, yogurt, and cheese. Dairy products play an important role in disseminating probiotic bacteria to humans because they create a good environment for their development and survival (Kim & Park, 2017). When employing probiotics in dairy products, the finished product's physical, chemical, and organoleptic qualities, the probiotic health impact, and regulatory and labeling requirements must all be considered.

Dairy drinks were the first foods to incorporate probiotics, and they have subsequently grown more popular than other probiotic beverages. Functional dairy drinks include whey-based beverages and fortified dairy drinks (which may contain probiotics, prebiotics, fibers, polyphenols, peptides, sterol, stanols, minerals, vitamins, and fish oil). The most popular probiotic bacteria utilized in the production of eight dairy probiotics drinks is *L. rhamnosus* GG. Because of its resilience to acid and bile, the *L. rhamnosus* GG probiotic is ideal for industrial application. More than a dozen commercial examples of dairy products containing probiotics reveal that lactobacilli, lactobacilli, lactobacilli, and *Lactobacillus plantarum* are the most often used probiotics.

According to a new study, probiotic bacteria can flourish in fermented milk based on several parameters. Numerous factors, including acidity and pH, dissolved oxygen and hydrogen peroxide concentrations, starter microbes, potential flavor compounds, and various additives (including preservatives), have been shown to affect the viability of probiotic bacteria during the fermentation and storage of fermented milk. Acidophilus milk, Sweet acidophilus milk, Nu-Trish A & B, Bifidus milk and Acidophilus buttermilk, Yakult, Procult, Gaio, and Prova are all dairy products that include probiotic bacteria. These bacteria also require the presence of certain chemical components, which milk lacks (Khoruts, 2018). A variety of milk components have been examined in an attempt to boost probiotic synthesis and survival in dairy products. The citrus fiber in fermented milk improves probiotic bacteria survival and proliferation.

The addition of soy germ powder to *L. reuteri*-fermented milk had favorable outcomes. Bioactive isoflavones generated during the fermentation of soy germ powder may protect *L. reuteri* against the detrimental effects of bile salts in the small intestine. A source of iron has been discovered in addition to whey protein, tryptone, yeast extract, and particular amino acids and nucleotide precursors. For probiotic bacteria to thrive in fermented milk, the strains of probiotic bacteria employed and the right manufacturing conditions must be used (both formulation characteristics and storage conditions). Yogurt, one of the original probiotic sources, is now a popular probiotic product. Yogurt has several dietary and health advantages. *Lactobacillus bulgaricus* and *Streptococcus salivarius subsp. thermophilus* strains are used to make yogurt. Lactobacilli and bifidobacteria can be added to the yogurt culture at any time.

There has been some debate concerning the involvement of these bacterial strains in the probiotic qualities of yogurt culture. It is crucial to assess probiotic survival and proteolytic effects in yogurt. The factors that influence the survivability of Lactobacillus and Bifidobacterium spp. bacteria in yogurt can be rather diverse. pH, hydrogen peroxide, dissolved oxygen concentrations, metabolite concentrations such as lactic acid and acetic acid, the medium's ability to function as a buffer, and storage temperature are all properties. The 9th Edition of Probiotics in Food and Agriculture Science states that although yogurt is widely used to provide probiotics, most commercial yogurts contain a small number of live cells (Jugan *et al.*, 2017).

Nutrition, growth stimulants and inhibitors, solute concentrations, inoculation amounts, incubation temperatures, fermentation times, and storage temperatures affect probiotic survival in yogurt. The strain has been discovered to negatively impact yogurt's probiotics. Probiotic organisms' viability can be ascribed to a decrease in medium pH and a buildup of organic acids due to growth and fermentation. The ultimate pH of yogurt appears to have the greatest impact on probiotic development and viability after fermentation. Organic acid metabolic byproducts may also influence probiotics' capacity to retain living cells during storage. The antibacterial properties of fruits and berries will likely interfere with the growth of probiotics in yogurt. To compensate for any potential loss of viability, high probiotic dosage may result in lower-than-expected product quality. The inclusion of probiotics has been proven to impact the yogurt's acidity, texture, taste, and appearance. Encapsulating yogurt probiotics with plain or coated alginate beads or chitosan-coated beads is one method for increasing the viability and stability of yogurt probiotics. Prebiotics and cysteine can also be added to yogurt to boost probiotic viability.

Cheese

The most prevalent dairy products containing probiotics are milk and yogurt. Cheddar or another sort of dairy product appears to be a superior choice. Cheeses outperform yogurt and fermented milk regarding pH and buffering capability, nutrients, solid consistency, fat content, and shelf life. According to several studies, probiotics may flourish in cheese even after it has passed its expiration date. After traveling through a simulated human GI tract, cheese probiotics were found to greatly increase the number of probiotic cells in the stomach. To give the same health advantages as yogurt, cheese requires a higher concentration and vitality of probiotic bacteria. Danisco conducted an industry-first study on cheese's potential to create and sustain healthy probiotic microbes in 2006. There were few alternatives for probiotic cheese at the time. According to the findings, fewer than 10% of the microorganisms were lost in the cheese whey (Johnston et al., 2012). Mills DA, an Oslo, Norway-based firm, created the first commercially sold probiotic cheese. On the commercial market today, fresh, semi-hard, and hard probiotic cheeses are available in over 200 different variants. If you want to add additional probiotics to your diet, look no further than semi-hard and hard cheese (about 4 to 5 times). Cottage cheese has a short shelf life and a high RDA when stored at room temperature. In other words, as a meal with tremendous potential as a probiotic carrier.

Milk-derived Products

Probiotics may be found in dairy products such as Quark, chocolate mousse, frozen fermented dairy desserts, sour cream, and ice cream. Two probiotic cultures were tested on Quark to see whether they might boost its nutritional content. According to a recent study, probiotics in skim milk may give the best mix of fat, protein, lactose, and phosphorus. Probiotic and prebiotic chocolate mousse has been made and evaluated. The probiotic chocolate mousse was made using *L. paracasei* subsp. *paracasei* LBC 82 and inulin.

The findings indicate that *L. paracasei* may be carried safely in chocolate mousse. Sour cream as a food carrier is effective as a probiotic carrier in studies. As a probiotic carrier food, ice cream has a lot of potential. Cruz and colleagues investigated the technology requirements for making probiotic ice cream (Hill, 2010). Several aspects must be considered, including culture selection, inoculum concentration, the optimal processing stage at which cultures are introduced, processing and transit temperatures, and storage temperatures.

Consequently, scientists discovered that probiotic cultures had no effect on the flavor or texture of ice cream or other frozen delicacies and that probiotics could survive in these goods for an extended length of time.

5.7 PROCESSED FRUIT-VEGETABLE

Fruit juice-derived probiotic products are becoming increasingly popular. Fruit juice's beneficial components make it a good vehicle for probiotic administration. Fruit juices are both pleasant and nutritious, with flavors that appeal to individuals of all ages. The fruits contain antioxidants, dietary fibers, minerals, and vitamins. According to a recent study, some demographic groups may be unable to take probiotics due to thirteen allergens. Probiotic strains with these qualities can be employed to make nutrient-dense fruit juice.

On the other hand, Probiotic bacteria's taste profiles will differ from those of regular, non-functional meals. After being infected with *L. plantarum*, orange juice, which is generally hated, took on a new flavor and aroma. As a result, when the health advantages are highlighted, individuals prefer them over conventional orange juice. Several attempts have been made to decrease the perception of unpleasant tastes and odors in probiotic fruit juice. According to Luckow and colleagues, probiotic off-tastes constitute a significant source of customer dissatisfaction. At the same time, tropical fruit juices like pineapple, mango, and passion fruit, which contain 10% (v/v), may be able to mask such flavors (Harzallah & Belhadj, 2013). Several research has been conducted on probiotic fruits. If noni juice is used as a raw material in synthesizing these probiotics, the probiotics *B. longum* and *L. Plantarum* may be the best choice for fermented noni juice. Many strains of bacteria were used to evaluate the fermented pomegranate juice including *L. acidophilus, L. Plantarum, L. acid arm, L. delbruekii*, and *L. paracasei*.

Pomegranate juice was shown to be a great probiotic beverage because it boosted the viability and development of the bacteria *L. plantarum* and *L. delbrueckii*. We are investigating the best circumstances for *L. casei* growth in cashew apple juice. *L. casei* survived in cashew apple juice for 42 days after being refrigerated. *L. casei* has been discovered to thrive in freezers, and cashew apple juice has been demonstrated to be an effective probiotic supplement. By fermenting banana puree with *L. acidophilus* immobilized in Ca-alginate, an enhanced banana probiotic with a higher number of viable cells may be created. *L. casei* must be immobilized on apple and quince pieces for a lengthy period to live in an acidic environment.

Fermenting vegetables has been done since the beginning of time. Probiotics can be consumed in a variety of ways, including fermented foods. The probiotic bacteria *L. acidophilus* and Bifidobacterium cannot be introduced into vegetable fermentation due to the low incubation temperature (Gwee *et al*., 2018). *Lactobacillus rhamnosus, Lactobacillus casei*, and *Lactobacillus plantarum* are better for fermenting vegetables. Despite this, probiotic bacteria on plant-based substrates proliferate rapidly when kept at 37 °C.

Much research has been conducted to discover novel probiotic veggies. The viability of using carrot juice as a source of Bifidobacterium strains for probiotic food manufacture was investigated. Even though there were no nutrients in the juice, bacteria in carrot juice were still able to execute metabolic functions. Tomato juice may be used to produce probiotics such as Lactobacillus acidophilus, *Lactobacillus plantarum, Lactobacillus casei,* and *Lactobacillus delbrueckei.* According to the researchers, the four LAB were able to rapidly produce cells and lactic acid from tomato juice without the need for additional nutrients or pH adjustment.

5.8 BAKERY

Because most probiotics are heat-sensitive, they are typically added to baked items after they have been cooked to preserve their viability. There are two ways to accomplish this goal:

- Probiotic spray-on and blend-in formulations may be applied to the product before packing. Probiotic formulations are used after baking, such as frosting, icing, or filling.

Probiotics may be absorbed in several ways; however including the probiotic into the dough during the manufacturing process offers the most benefit. This strategy would be great for avoiding the penultimate stage and ensuring an equal distribution of microorganisms. Thanks to this technology, probiotics may now be added to a broader range of baked goods, including savory and sweet ones (Gogineni *et al.*, 2013).

Before they may be used in baking, probiotics must fulfill several criteria. Microbes must be safe, stable, and not interfere with the intrinsic properties of the product while being employed in manufacturing and over the product's whole shelf life.

5.9 PET FOOD AND LIVESTOCK INDUSTRY

Probiotics are live microbial feed additives that help balance the host animal's intestinal flora . They have been awarded approval to provide several benefits to the host animal and the production of animal products. For instance, they are applied as animal feed in poultry, ruminant, pig, and aquaculture to increase animal health and food safety. The bacteria found in the digestive tracts of poultry are crucial for good digestion and animal health. Certain feed additives greatly affect this microbial ecosystem and its health advantages. Finding alternatives to antibiotics in chicken production is gaining popularity due to recent issues in several nations over the unforeseen severe adverse effects of antibiotics.

To deliver meals of similar quality and safety, probiotics have been advocated as a feasible natural alternative to antibiotics in chicken production. In addition, it has been demonstrated that adding probiotics to chicken feed boosts the interior and exterior quality of eggs. Adding probiotics to chicken feed enhanced egg weight, specific gravity, albumen, and shell weight, as well as the form index (Gupta, 2011). Environmental pressures are constantly exerted on farm animals, which might change their gut ecology and increase their sensitivity to pathogen infections. Introducing probiotics in animal feed has lessened the frequency of illnesses in farm animals. The administration of probiotic LAB and yeast to Holstein's calves enhanced their growth and lowered the frequency of diarrhea.

Applications of probiotics and prebiotics in animal nutrition contribute to the formation of safe food Probiotics have been exploited to treat gastrointestinal sickness in pigs and lessen pathogen loads. In addition to the in vitro test, several *in vivo* experiments utilizing a spectrum of probiotic bacteria are done to discover the most promising probiotics. In most experiments, increasing the number of helpful bacteria, reducing the burden of pathogens, enhancing immune cell responses to pathogens relative to controls, and strengthening resistance to pathogen invasion were found to have favorable benefits. Other writers indicated that the sickness was hurried or that the diarrhea was partially eased.

Due to their propensity to bind to intestinal mucus and create antimicrobial metabolites, probiotics are frequently exploited in aquaculture. It was shown, however, that the introduction of probiotic strains (obtained from the clownfish, *Amphiprion percula*) can prevent the attachment of these microorganisms to fish intestinal mucus and compete with the pathogens (Gao *et al.*, 2022). Common fish infections include *Aeromonas hydrophila* and *Vibrio alginolyticus*. The illness generated by *Vibrio parahaemolyticus* can be minimized in shrimp by feeding probiotics.

CHAPTER 6

PROBIOTICS LEGISLATION AND SAFETY CONCERNS

CONTENTS

Probiotics were defined as "living microorganisms that, when eaten in adequate numbers, impart a health benefit on the host" in the consultation. Biotherapeutic drugs and beneficial microorganisms that aren't found in food were thus left out. Instead, Lactobacillus and Bifidobacterium members could be detected as probiotics in food that were the focus of the study (Faujdar *et al.,* 2016). There were no specific concerns about genetically modified organisms, but the ideas and criteria applied to all probiotics were acceptable

Product demands, quality control, regulatory issues, and the therapeutic impacts of probiotics on human health, were all discussed in detail.

When it comes to being used in food, probiotics have to meet several requirements, including being able to multiply in the stomach and enduring passage through the digestive tract. Thus, they must be capable of thriving in bile-rich intestines or provided a diet that shields them from bile exposure while passing through the stomach and intestine. If probiotics are to affect the host, they must be able to grow and/or operate in the human body (Ezema, 2013). The uniqueness of the activity, not its microbial source, matters most. There must be an investigation of the viability of each strain in the chosen location.

To guarantee worldwide comprehension, the process of classifying and naming Probiotics should be referred to utilizing the International Code of Nomenclature. To ensure total openness, it was recommended that probiotic strains be included in a world-renowned culture collection. Because probiotic characteristics vary by strain, methods for strain identification, such as PFGE DNA/DNA hybridization and 16S RNA sequencing, have been proposed.

6.1 IDENTIFYING AND MEASURING THE BENEFITS OF HEALTH CARE

One of the several health advantages of probiotics is a reduction in acute diarrhea caused by viruses and pathogens, diarrhea induced by antibiotics, and diarrhea caused by *Clostridium difficile* and infection. Probiotics may also be useful in treating inflammatory bowel disease and some cancers; however further study is needed. Other established benefits include allergy reduction and immunological stimulation. Probiotics may be able to alleviate gastrointestinal inflammation by influencing the microbiota in the gut, which has been linked to a variety of diseases.

6.1.1 Methods for Evaluating Health Benefits

To assess the potential health benefits of probiotics, *in vitro* research was agreed upon before doing in vivo tests. Testing for acid and bile tolerance, antimicrobial production, and intestinal cell adhesion should be undertaken depending on the health benefit. To assess the efficacy of the probiotic product *in vivo*, randomized, double-blind, placebo-controlled human trials are required. When evaluating the health benefit of human research, it is crucial to incorporate clinically relevant data from the study population (Deng *et al.*, 2020). In certain nations, the objective of diarrhea research is to lower mortality rates. In others, it may be to avoid significant weight loss, shorten the disease duration, or accelerate recovery time.

6.1.2 The Capacity to Grasp How Systems Function

The chemical pathways via which probiotics benefit the body are unknown. This is a perplexing subject of study since the processes involved may vary from probiotic to probiotic. Probiotics, for example, produce an enzyme (-galactosidase) that hydrolyzes lactose into its constituent sugars, alleviating symptoms of lactose intolerance. Experiments conducted *in vitro* and *in vivo* should be designed at the molecular level to understand better how probiotics operate.

6.2 FEARS FOR ONE'S SAFETY

Standardized testing has been demonstrated to be required for evaluating the drug sensitivity or resistance profiles of Lactobacilli and Bifidobacteria. Another area that requires further research is the potential transmission of genetic elements to other intestinal/food-borne microorganisms, given the lack of knowledge about the conditions under which these genetic elements could be mobilized and the uncertainty about whether these conditions could lead to clinical issues. As discussed further in this article, it has been concluded that standards describing essential requirements are required for human safety evaluation (Deshpande et al., 2011).

All nations' governments have yet to recognize probiotics as a food ingredient. Probiotics are mostly classified as meals and dietary supplements, which means they can only make broad health claims.

Probiotic foods, particularly powdered milk products, must be produced, handled, and kept so that their viability and probiotic activity are maintained. Probiotic microorganisms in food must be precisely quantified to ensure their inclusion on product labels. Prebiotics have been recognized for their potential benefits when combined with probiotics and their ability to boost natural beneficial bacteria in the body. Finally, it was recognized that a rigorous trace-back system is essential for evaluating the health advantages of long-term probiotic medication (as well as any possible side effects).

Finally, the experts reached the following conclusion: Adequate scientific evidence supports the premise that eating probiotic-rich foods can promote health. According to the requirements for assessing probiotics, greater research is needed to establish the health benefits of probiotics to people. There is strong evidence that certain probiotic strains are safe for human ingestion and can provide specific health benefits to the host. These advantages, however, cannot be applied to other strains without further study. Among the health benefits that can be obtained are allergy relief and stomach and urogenital infections. The medical profession should research probiotics to cure and prevent these illnesses. According to a recent study, probiotics may help healthy people prevent some diseases and modify their immune systems (Di Cerbo & Palmieri, 2015). Probiotics are regulated differently worldwide, and as a result, specific health benefits cannot be listed on probiotic product labels.

These suggestions were made: The International Code of Nomenclature (ICN) must be observed while identifying and naming possible probiotic strains. To be called a probiotic, the bacteria in the actual product vehicle must be capable of conferring health benefits on the host. More in vitro and animal investigations are required. More statistically significant proof of human effectiveness is required. The use of quality assurance procedures, the establishment of shelf-life restrictions, and clearly labeling the least dose and health claims that may be substantiated are all required. In terms of legislation, there must be universal agreement on the position of probiotics as food components. A regulatory framework must be built to effectively address probiotic problems such as efficacy, safety, labeling, fraud, and claims. Probiotic products should have the ability to detect particular health benefits. To track long-term health benefits and record and analyze adverse events, probiotics in food should be tracked using trace-back and post-marketing protocols (Daliri & Lee, 2015). Probiotic products should be available to a broader number of individuals, particularly those at high risk of disease or death or in the humanitarian sector. Probiotic criteria and processes still need to be researched further.

The FAO website has a comprehensive copy of the FAO/WHO study from 2001. The paper included general principles for evaluating these probiotics.

Specific issues concerning their pathogenicity, toxicity, allergenicity or other related nutritional attributes or safety characteristics were addressed. Scientific assistance on probiotics' nutritional and safety assessment was also provided to FAO, WHO, and their member countries.

A collection of individuals

When asserting that a meal has a probiotic impact, it is advised that the study's criteria be followed. Probiotic certification requires the following basic requirements: strain identification, *in vitro* testing to screen potential probiotics, safety evaluation, and in vivo research to validate effects.

Identifying probiotic strains by genus and species is critical for epidemiological studies and correlating a specific strain to specific health effects. Probiotic strains must be stored in a reputable culture repository. In vitro testing of probiotic bacteria, like animal research, is critical for assessing their safety and their functional properties and modes of action. However, because the current in vitro approaches cannot predict the action of probiotic bacteria in the human body, the efficacy of probiotics for human usage must be proved through human studies (Attri *et al*., 2021).

Probiotic strains should be described at the very least utilizing the following tests, especially in a group of bacteria that are widely considered safe: Specific metabolic processes such as D-lactate production and bile salt deconjugation are evaluated, as are adverse effects on humans, epidemiological surveillance of adverse consumer events, and testing for toxin production (if the strain under investigation comes from a species known to produce mammalian toxin) and hemolytic activity.

Immunocompromised animals may be used to examine a probiotic strain's capacity to withstand infection.

There should be phase 2 (efficacy) research on probiotic foods that are double-blind, randomized, and placebo-controlled (where the placebo is the food carrier devoid of the test probiotic). Statistical significance must be applied to physiologically relevant data when estimating sample size for defined purposes.

Human trials should be the major source of evidence on probiotic efficacy to discover advantages such as enhanced health, well-being, quality of life, decreased disease risk, a long time between episodes, or faster recovery from sickness. It is recommended that the findings of human studies be published in peer-reviewed scientific or medical journals to assure the validity of the findings (Dai et al., 2018). When diet is considered, probiotic supplementation should have no negative side effects.

To provide the advertised health advantages, an appropriate dosage of probiotic bacteria must be present in the prescribed serving size. The necessary storage conditions should also be indicated on labels. Labels should also specify the minimum viable numbers of each strain of probiotic bacteria, and suggested serving quantities and the purported health benefits.

The three recommendations given in the report are as follows:

Adoption of probiotics as living bacteria that enhance a host's health when given in adequate quantities.

Bacteria must fulfill the conditions outlined in this study before they may be termed "probiotic." There should be a regulatory framework that permits probiotic food labels to make specific health claims when scientific evidence supports those (Chugh & Kamal-Eldin, 2020).

The fourth stage is to make these suggestions available all across the world.

Probiotic foods must be manufactured in line with established manufacturing techniques and under proper storage conditions.

6.3 CODEX AND HEALTH AND NUTRITION CLAIMS

The Codex Alimentarius Commission was founded in 1963 by the Joint FAO/WHO Food Standards Program to develop food standards, guidelines, and other relevant literature, such as codes of practice. There are now 163 member nations. The fundamental aims of this program are to protect consumer health, ensure fair trade practices in the food business, and stimulate international collaboration creating food standards.

The Consultation on Probiotics in Foods and the Guidelines for the Evaluation of Probiotics in Foods have submitted results and suggestions to the Codex Committee on Food Labeling (CCFL) and the Codex Committee on Nutrition and Foods for Special Dietary Uses (CCNFSDU). The CCFL governs prepackaged food labels, nutritional labeling requirements, and nutrition claim recommendations. The CCFL aims to create globally applicable labeling regulations (Choudhary & Tandon, 2009). Other objectives include examining particular Codex Commission concerns and researching food advertising difficulties, emphasizing deceptive claims and descriptions. The Codex Committee on Labeling is in charge of all of these tasks.

The mission of CCNFSDU is to conduct nutritional research, advise the Codex Alimentarius Commission on nutrition issues, draft general provisions on nutritional aspects of all foods, develop standards and related texts for foods for special dietary use, and consider, amend, and support proposals to include nutritional aspects in Codex standards and related texts. Neither food nor its preparation should be presented so that its genuine nature is misrepresented in any way. When it comes to claims, all foods are subject to the same standards. A claim must be supported by evidence to be judged valid; some of these claims fall under international law. Currently, most nations allow only broad health claims on food labels that do not expressly specify illness prevention, alleviation, or therapy. Specific health claims should be acknowledged when there is adequate scientific evidence to back them, according to the Working Group Probiotic Guidelines and the FAO/WHO Consultation (preferably supported by an independent third-party review by scientific experts). The Probiotic Guidelines are hoped to be utilized as a scientific review of new food in the Codex draft on health and nutrition claims (Chalas *et al.*, 2016). The probiotic organizations' work in this context for scientific examination and evidence of probiotic health claims should result in a model for Codex draft guidelines, health and dietary claims, and particular reference to probiotics in the Codex draft.

6.4 FDA REGULATIONS FOR PROBIOTICS

FDA is developing an improved methodology for determining the purity of probiotic products

Researchers from the US Food and Drug Administration provided proof of concept for identifying possible probiotic toxins(FDA).

It eliminates beneficial microorganisms from a product while enabling potentially harmful contaminants to proliferate unchecked, allowing regular laboratory processes to detect potentially hazardous pollutants.

Probiotics and other live biotherapeutic therapies are manufactured from naturally occurring microorganisms such as bacteria (LBP). The most frequent bacteria are Lactobacillus and Bifidobacterium. Suppose the probiotic is intended to diagnose, treat, mitigate, or prevent a human ailment. In that case, an Investigational New Drug application must be filed to the FDA's Center for Biologics Evaluation and Research to conduct clinical trials on human volunteers.

Despite the common opinion that probiotics sold over the counter are safe, clinical research on persons with compromised immune systems, such as those with a disease or who have taken immunosuppressive drugs, may be conducted.

Over-the-counter probiotics utilized in clinical studies to examine their potential value for various health conditions must fulfill more stringent quality requirements to confirm their purity and efficacy. Microbial purity, or the absence of undesirable germs, is one of the most important safety criteria for LBPs in clinical studies (Culligan *et al*., 2009). Probiotic bacteria may reduce detection sensitivity, making it more difficult to spot hazardous microbial contamination. This is correct. One method being researched to improve the sensitivity of detecting tainted microorganisms is restricting or inhibiting the development of product bacteria. FDA scientists created recombinant phage lysins that were used in improved LBP purity testing. Rapid and dependable testing for identifying dangerous contaminants that may enter LBPs via various manufacturing processes or product handling is crucial.

When FDA researchers sought to conduct "mock" purity testing on *Lactobacillus jensenii*, they generated a lysin, a recombinantly manufactured enzyme that imitated the probiotic product strain. Certain viruses, known as "bacteriophages," manufacture their form of "lysin" to destroy infected bacteria and allow newly produced viruses to propagate. The researchers developed a lysin against *Lactobacillus jensenii* cells to eliminate them from a culture that had been "spiked" with modest numbers of organisms suggesting possible bacterial pathogen contamination (either *Escherichia coli* or *Staphylococcus aureus*).

Figure 6.7: Probiotic drinks.

When *L. jensenii* was decreased or eliminated, it was simpler to detect *E. Coli* or *S. aureus* contamination at lower levels. Before doing an LBP test, it is necessary to eliminate all of a probiotic product's beneficial bacteria (Cammarota *et al.,* 2014).

It is believed that by utilizing the FDA study as a foundation, researchers can to develop LBP purity testing reagents and devices more swiftly in the future.

6.5 REQUIREMENTS OF PROBIOTICS

The FAO/WHO definition of probiotics may be converted into four basic criteria for determining if a certain strain of bacteria qualifies as a probiotic for inclusion in meals and supplements. It is safe for the intended use, supported by at least one successful human clinical trial conducted following generally accepted scientific standards or, when applicable, following local/national authorities' recommendations and regulations, and alive in the product at an effective dose for the duration of the shelf life. Here, we go over how to examine each of the four criteria above in great detail. These guidelines must be widely adopted for the term "probiotic" to be accurately used in scientific articles, commercial labeling, and discussions with regulators and the general public.

In this context, the terms "probiotic," "criteria," and "living microbes" are all significant. Identification of species, safety, and health advantages of lichen and Bifidobacterium

Consumers are getting increasingly interested in nutritional supplements and food-based health care. A growing trend is using evidence-based treatments to enhance diets and lives. As a result, the market for probiotic-containing meals and supplements has become increasingly broad. During an expert meeting held in 2001 under the auspices of the World Health Organization (WHO) and the Food and Agriculture Organization, probiotics were defined as "living microorganisms that, when administered in suitable proportions, exert a health effect on the host" (FAO).

Probiotics must have "defined contents, an appropriate viable count at the end of shelf life, and adequate demonstration of health effects," In a policy statement released in 2018, the International Scientific Association of Probiotics and Prebiotics (ISAPP) confirmed these findings. International Society for Professional Appraisal Advancement (ISAPP, 2018). Despite widespread misinterpretation, the term "probiotic" is still widely used in the food and supplement industries. This section, defines the minimum requirements for using the term "probiotic" correctly (Charalampopoulos & Rastall, 2009). When new "biotic" names reach the global language, this is a very good time to use this (e.g., pharmabiotic, psychobiotic, postbiotic, synbiotic, and others).

The definition of probiotics may be reduced to a few basic factors. Several probiotic companies have made it a goal to develop a set of these criteria. If a proposed strain or strain combination fits the criteria for probiotic certification, it will be shown as a decision tree in Figure 1. This decision tree may be used for anything. A collection of "commandments" or a "checklist," such as the one shown in this ISAPP infographic, can also be employed (2017).

Using a decision tree, assess whether a potential probiotic meets the definitional requirements.

Traditional probiotics are not the only ones included by the word "probiotic." As a result of technical breakthroughs, isolating probiotics from previously undiscovered sources with previously unknown health benefits will be necessary. This does not exclude using "probiotics" in the future, which may be considered live biotherapeutics in some situations. Probiotics, however, must be manufactured under appropriate safety, legal, and ethical issues, depending on their intended purpose. Before removing germs from humans, the Nagoya protocol must be followed, and informed consent obtained.

These criteria must be defined explicitly and per previous publications without delving into the complexities of any hypothetical mechanistic requirements. Probiotic strains must quickly satisfy the following requirements: A product will remain on the market for the duration of its shelf life if it meets all four criteria outlined above: adequate characterization, safety for the intended use, and the result of at least one successful human clinical trial conducted by generally accepted scientific practices (Bowe *et al.*, 2014).

Probiotic that is adequately described as-

The most crucial parts of good probiotic characterization are probiotic strain identification and naming. Probiotic strains should be identified according to the International Code of Nomenclature, in conformity with currently accepted bacterial nomenclature.

It is important to demonstrate that a strain belongs to a well-known genus, species, and subspecies to identify probiotic bacteria. As a result, strain typing is critical for determining if probiotic effects are strain-specific. Nomenclatural rules require that the name of a strain include the genus, species, and subspecies names, as well as the catalog number of the strain's culture collection or other commercially accessible identification. Because the strain can be utilized for research but not necessarily for commercial reasons, we recommend that it be donated to a reputable culture collection (Ayichew *et al.*, 2017). As a general guideline, do not use several strain names for the same organism. To ensure that their goods contain the same strain throughout time, manufacturers must ensure that their strains are genetically pure.

Since phenotypic techniques are no longer sufficient for trustworthy identification, various molecular technologies have favored the use. Identification strategies may alter in the future depending on the organism being recognized. If valid reference sequences are employed, 16S ribosomal DNA sequencing is a well-known and reliable approach for species identification. The generated sequence may be compared to massive reference databases encompassing virtually all known bacterial varieties; however, curated databases like PATRIC are preferable for final confirmation. If there is any uncertainty or a complete identification that integrates additional molecular and phenotypic criteria is required.

The "gold standard" for strain identification is whole genome sequencing (WGS), which includes any extrachromosomal components. Using a correctly sequenced genome, microorganisms may be recognized at the species and strain level. NCBI, for example, has a collection of reference databases. One of the numerous benefits of whole genome sequencing is the capacity to identify strains precisely and consistently and check for risk factors (for more safety information, see the section "Probiotic is Safe for Intended Use"). This study may make it simpler to detect plasmids, which are ubiquitous in some lactic acid bacteria and may be required for the probiotic effect. Other types of typing approaches, in addition to WGS, may be employed to compare strains. Even though de novo species or genus identification is not achievable, strains can be compared using multilocus sequence typing or pulsed-field gel electrophoresis.

In an ideal world, identification would be performed by specialized or suitably qualified labs with access to the appropriate reference databases and methodologies and technologies that are updated, confirmed and calibrated. Because any strain might be regarded as a probiotic in probiotic combinations, it is critical to precisely identify each strain (Zhang *et al.*,

2016). A strain's probiotic activity should be validated by its characterization. It should be noted, however, that clinical data is necessary to substantiate a claim of probiotic efficacy. Nonetheless, testing for important probiotic characteristics may uncover plausible processes behind clinical results.

These features include the ability to persist in crucial bodily locations, produce short-chain fatty acids such as lactic acid, adhere to mucus or intestinal epithelial cells, and interact with human immune cells. The Belgian Superior Health Council's documentation on how to identify, strain type, and evaluate the safety of probiotics is quite useful. It should be noted, however, that the existence of probiotics is not dependent on the presence of these phenotypic features (Yao *et al.*, 2020). These assays may provide insight into a probiotic's efficacy or be useful in the screening phases, but they have not been proven to be accurate indications of probiotic efficacy in humans. The mere presence of one or more of these genotypes or features in a strain does not qualify it as a "probiotic."

According to research, the highest association between probiotic strains and certain health outcomes may be seen in the evidence pool. The health benefits of a specific strain or taxonomic group are not necessarily confined to that strain. This assignment requires just one probiotic microbial characteristic. Consider the circumstance in which the enzyme mediates the symptoms of lactose intolerance. This feature may be species-specific rather than strain-specific. Even if a common trait helps, identifying probiotics down to the strain level is impossible.

6.6 SAFETY OF PROBIOTICS

Food and dietary supplement manufacturers are obligated to guarantee that their goods are safe for customers. The first step in determining a probiotic's safety for inclusion in meals and dietary supplements is to identify it down to the strain level precisely . Historical data or trials are utilized to demonstrate the product's safety. A product's overall degree of safety is determined by its history of safe use. To assess if a product is safe, science-based criteria must be employed, which includes doing an appropriate phase 1 study (Wescombe *et al*., 2012). A probiotic strain must adhere to the national/regional authority's safety criteria, as described above. There are no defined requirements for adequate proof in the case of strain safety, and the review is done on a case-by-case basis.

6.6.1 Human-Safe Probiotic Species and Strains Must Exist

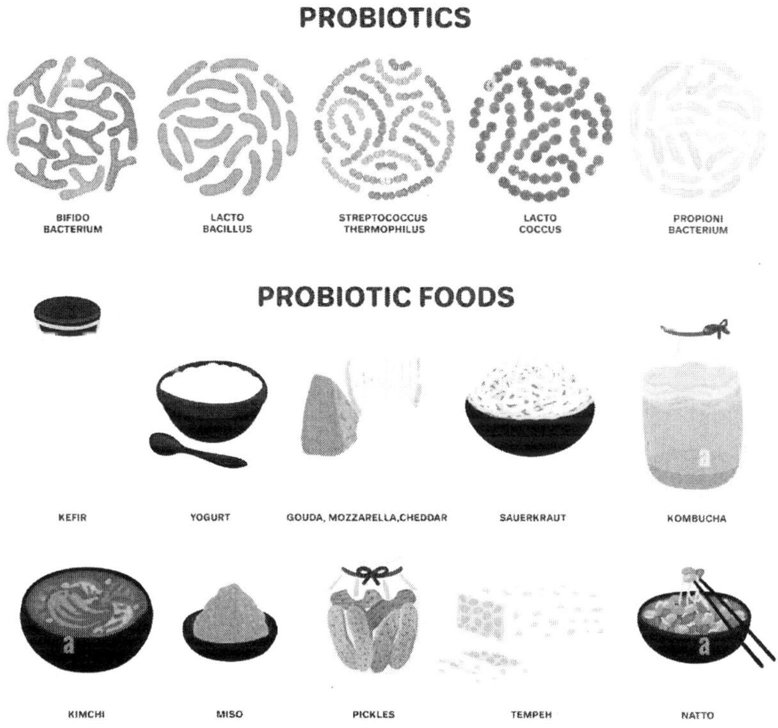

Figure 6.1: probiotics and their food.

Concerns concerning the safety of injecting living bacteria must be addressed before probiotics may be used frequently by the healthy population. Many commercially available probiotic strains, including lactic acid bacteria, Bifidobacteria, and yeasts, have been proven safe for use in food or dietary supplement formulations. They are from genera and species with a long history of being used safely as starter cultures or probiotics. Under the "Qualified Presumption of Safety" (QPS) concept, the European Food Safety Authority (EFSA) has maintained a list of species considered safe for human consumption in foods since 2007.

6.7 THE EUROPEAN FOOD SAFETY AUTHORITY

The QPS technology offers a rigorous, evidence-based, and often updated strategy to relay information on the safety of certain microorganisms. When food strains are members of a QPS species, the list can be used to establish their safety, as long as the strain-specific tests described below are also completed. This list was created using research on a wide range of microorganisms often identified in the food supply chain and historical data and current knowledge monitoring (EFSA, 2020b). The QPS does not expressly examine the risks of probiotics for vulnerable populations or other applications of probiotics other than food; rather, it focuses on the use of probiotics by the general, healthy population. QPS is insufficient since depends on EFSA submissions for pre-market clearance in the EU market. Many microorganisms used in traditionally fermented foods are not on the QPS list. The European Union Novel Food Regulation may apply to strains not QPS species before they may be sold in Europe. In the United States, for example, the generally recognized as safe (GRAS) criterion is used to assess the safety of probiotics.

6.7.1 Probiotic Species and Strains Must Fulfill Strict Safety Requirements

Species identification is the cornerstone of every safety evaluation, according to the principles given in Section "Introduction." Risk factors particular to a species or genus must also be identified and examined at the strain level. The lack of virulence factors or genes associated with antibiotic resistance is the most serious. The EFSA has defined the EU's antibiotic resistance and susceptibility phenotypic cut-off values. Use these criteria to assess yeast, bacterial, and antimycotic antibiotic susceptibility. Techniques for screening the phenotypes of potential bacterial strains are well-established. If resistance is seen above the cut-off values, a more thorough explanation is required (Valero-Cases *et al.*, 2020). Based on WGS results, genes associated with claimed resistance may or may not be present. When resistance genes are discovered, they must be examined for transposable elements. This indicates that if this is the case, it cannot be ruled out that the resistance gene is transferrable, so marketing the strain is not advised.

On the other hand, the genome sequence may be used to search at least two databases for antibiotic resistance genes. A Hidden-Markov model database should be employed for microorganisms that are underrepresented in existing databases. The genome may need to be studied for genes coding for recognized virulence factors like toxins, invasion, and adhesion factors based on the strain's classification and intended purpose. If antibiotic resistance cut-off values are unknown, the producer must confirm that the proposed probiotic strains do not include genes that may be passed on to other organisms and that their resistance profile matches that of other individuals of the same species. To guarantee that susceptibility testing techniques are appropriate for and tailored to the microorganisms under consideration, additional data on the susceptibility profiles of the taxon in issue may be required.

Phenotypic features such as the ability to produce biogenic amines and d-lactate may be examined for safety at the strain level (Tadesse, 2012). Both can be checked using widely accessible genetic analysis or recognized phenotypic assays. Additional strain-level tests of hemolytic and bile salt hydrolase activity are performed; however, it is unknown if these measurements affect safety.

6.7.2 Vibrational Safety Testing in Depth

According to the European Union's view that needless animal research should be avoided for ethical and practical reasons, in vivo safety tests for the bulk of present probiotics, which are QPS species with a proven history of safe usage in foods, are questionable. QPS species should have little to no effect on healthy animal models such as mice and rats.

By methodically documenting and reporting adverse events, probiotic safety and tolerance may be reliably documented in human intervention studies. Vital signs and other biological and clinical indicators can be utilized to gather critical safety information. There may be a safety issue if there are unanticipated deviations from the norm or baseline. These safety results must be documented, assessed, and reported following recognized scientific methodologies for human research. Sponsors, researchers, authors, and journal editors should collaborate to make disclosing safety and tolerance data easier in probiotic human clinical studies (Smolyansky, 2010). No matter how long or how many individuals are engaged, every study is certain to produce some bad results. It is necessary to research whether adverse occurrences differ or are considered intervention-related between intervention groups (probiotic or placebo). Probiotic-related adverse effects have only been observed seldom, slightly, and briefly in investigations with healthy participants. Probiotics should be administered under medical supervision in some sensitive groups (such as the elderly, pregnant women, and those with weaker immune systems).

A huge variety of Lactobacillus, Bifidobacterium, and yeast strains found in foods and nutritional supplements for the general population offer no significant safety concerns due to their long history of safe use and thorough studies and monitoring. The major focus is on the safety of food and dietary supplements; however, other uses may have different safety standards, for example, depending on how or at what dose the product is supplied.

According to generally accepted scientific criteria, at least one human clinical study has been conducted to support the probiotic. Since 2001 and again in 2014, the ability to confer health benefits on its host has been an essential part of the definition of a probiotic. Using probiotics, for example, has a positive impact on a person's health, which we call a "health benefit." This text is meant to inspire exploration of a wide range of health-related topics because it is not prescriptive. For the suggested microbial strain, one or more human studies demonstrating a health benefit are required (s). Following that, a confirmation trial would be preferable (s). This criterion is qualified because the trial must be conducted following accepted scientific practices (Abuajah *et al.*, 2015). Some authorities believe that strains of a species (or another taxonomic group) can legitimately use the term "probiotic" when it has been demonstrated that many members of that species bestow a benefit due to a common mechanism.

Lactase activity in strains like *Streptococcus thermophilus* and *Lactobacillus delbrueckii subsp. bulgaricus* can help alleviate lactose intolerance symptoms. Because of their beneficial effects on the gastrointestinal tract, probiotic strains of *S. thermophilus* and *L. bulgaricus* may be referred to as such. To use the term "probiotic" to describe these strains of microorganisms, probiotics must be discovered and proven to have the required trait. The validity of a health benefit demonstration is affected by the quality and soundness of the trial itself (i.e., how well it was planned and carried out) and the scientific community's ability to critically evaluate published trial results (i.e., how well it was reported). Various strategies for designing and reporting probiotic clinical trials, as well as risk of bias (RoB) assessment and critical review, may be beneficial.

Tools to aid in the design, reporting, bias risk management, and critical review of human probiotic intervention studies.

6.8 CONSIDERATIONS FOR PROTOCOL DESIGN

Two examples of internationally recognized guidelines for the clinical trial design are the International Council for Harmonization (ICH-GCP) and country-specific, legally binding versions of its Good Clinical Practice guidelines (GCP) (and conduct). Initially, these guidelines were implemented to ensure the safety and integrity of the study's participants. Aside from the blatant ethical benefits, adhering to ICH-GCP criteria helps to produce better, more reliable data. The International Committee of Medical Journal Editors (ICMJE) and the World Health Organization (WHO) have collaborated to develop an internationally recognized tool for developing clinical trial protocols that follow their guidelines.

This tool is supported by journals, funders, regulators, and academic institutions worldwide (ICMJE). The "Standard Protocol Items: Recommendations for Interventional Trials (SPIRIT)" checklist from 2013 includes 33 components that must be included in all clinical trial protocols. The protocol should be published in a public database (such as ClinicalTrials.gov) before the start of the study. This, in theory, should encourage the development of higher-quality studies and more open reporting of findings (Stavropoulou & Bezirtzoglou, 2020). Another good practice that should be promoted in the probiotics industry is the publication of clinical trial protocols in peer-reviewed journals, which often require that the protocols be recorded in a public registry.

Several design flaws plague probiotics research, effectively ruling out formal inferences of a health benefit. Some examples of study design information that may be included are participant allocation concealment, blinding (double-blind versus single-blind or open-label), control selection (placebo versus comparator treatment), dosing, and administration regimens (concentration, administration schedule, start and duration of supplementation) (health status, age, and gender). Because of the inherent individuality of potential microbial strains, there are no general probiotics field guidelines that favor one research design type over another or one dose protocol over another. However, during the research design stage, a thorough evaluation of these variables, and the microbial strain and the characteristics of the target population, is required. Learning about existing reporting guidelines and resources for critical evaluation of published trials may aid in achieving this goal.

6.9 REPORTING TRIALS CORRECTLY AND ACCURATELY

CONSORT is a set of guidelines that authors must adhere to when publishing the results of their trials various international journals. CONSORT has emerged as the de facto standard for reporting and disseminating trial data. CONSORT was not intended to be used as a guide for conducting or designing trials. However, it may be useful to understand the aspects that must be reported before a trial. The SPIRIT 2013 protocol design guidelines prioritized transitioning from a SPIRIT-compliant procedure to a CONSORT-compliant report.

If trials are reported by the CONSORT 2010 standards, researchers will be able to critically analyze their findings more easily in the future, leading to stronger conclusions in meta-analyses and systematic reviews. CONSORT, in a nutshell, covers all aspects of trial design, data collection, analysis, and reporting. It is a requirement of CONSORT, for example, that a participant flowchart explicitly states the number of participants for each stage of the study, including recruitment, retention, and completion of the study and the exclusions, losses to follow-up, and populations to treat or per protocol. This information is crucial for future quality and RoB assessments (e.g., systematic reviews for evidence-based medicine guidelines or regulatory purposes). CONSORT also requires that authors explain any biases or uncertainties that may affect how their findings should be interpreted in their papers (Aggarwal *et al.,* 2013). It is expected that the results will be described in generalizability and that the health risks and benefits will be clearly understood (implying a detailed reporting of the adverse events) (implying a detailed reporting of the adverse events).

6.10 CRITICAL APPRAISAL AND RISK OF BIAS ASSESSMENT OF PUBLISHED TRIALS

Clinical study critical evaluation and RoB assessment are critical components of evidence-based medicine. They allow for the objective evaluation of the weight of a trial's findings. Numerous methods have been created for these reasons, most of which are intended for authors of systematic reviews and meta-analyses or best practice recommendations but can also be useful for reviewing the quality of trials published without using a reporting standard such as CONSORT 2010. This applies to trials published before 2010 and a handful of more recent trials. Generally, compliance with the CONSORT 2010 standards remains low in the medical literature.

Although the distinction between critical evaluation (i.e., quality assessment) tools and RoB assessment tools may appear confusing, the two methodologies are separate. For example, the RoB evaluation methods used by Cochrane review writers are intended to determine if the trial's results are free of bias and reliable Quality assessment tools, on the other hand, frequently include parameters relating to both reporting quality (e.g., obtaining ethical approval or describing power calculations) and research quality, transparency, and consistency (e.g., randomization and allocation concealment, proper control selection, and missing outcome data). The latter is closely tied to probable sources of bias identified by RoB techniques.

In general, the weight given to trial outcomes is proportionate to how effectively sources of bias were. Given that lack of randomization, blinding, or controls have been identified as significant sources of bias in clinical trials, double-blind RCTs have emerged as the "gold standard" design for reliably demonstrating health benefits, as evidenced by the higher score assigned to RCTs over other designs when grading the quality of evidence from clinical trials. Other research designs, such as open-label and uncontrolled investigations, are feasible and have been utilized in the past. While the results of such research may not be substantial enough to justify a probiotic classification on their own, they can provide helpful supporting material.

While the first three requirements apply to a microbial strain to be classified as a probiotic, the fourth criterion relates to the product that contains the probiotics. The definition of probiotics does not specify a dose but rather specifies that probiotics should be delivered in levels sufficient to provide a health benefit to the host. Given that probiotics are live bacteria capable of self-replication inside the host, it is plausible that a few probiotic cells might be sufficient to elicit a favorable impact over time, provided they multiply sufficiently within the host (Saraf *et al.*, 2010). This is especially true for pathogenic microorganisms that cause disease and can hurt host health at shallow levels due to their virulence and ability to proliferate within the host.

Dose range studies were designed to establish an active agent's acceptability, effectiveness, and safety profile that can only be administered in fixed concentrations and cannot multiply after delivery. As a result, dosage range studies are popular in clinical trials but are less common in food and dietary supplement trials. This is primarily due to the safe assumption for food additives. Dose-ranging studies are often undertaken in clinical settings when the maximum tolerated dose (MTD) for the bioactive under investigation has been determined. We are unaware of any human oral MTD studies for probiotic strain or strain combination (section "Probiotic is Sufficiently Characterized"). Given the outstanding safety profile of probiotics and the fact that they have seldom been submitted to MTD or dose-ranging trials, most investigations simply chose a daily dosage between 108 and 1011 colony forming units (CFU), which mirrors effective levels in previous research. While establishing an optimal dose that results in a certain health benefit may be of interest, it is not a requirement.

Probiotic strains' viability should be quantified using standardized enumeration methods such as plating; CFU counting on selective growth media, such as Lactobacillus acidophilus and Bifidobacterium or flow. Several publications have evaluated the methods. The vitality of probiotic strains at the effective dose should be verified in test products during clinical research and guaranteed in commercial goods until the end of shelf life according to quality norms.

Culture-independent metagenomics approaches based on high-throughput next-generation sequencing have been developed to quantify probiotic combinations; however, these methods do not guarantee that what is tallied is viable. These approaches can provide useful information about possible pollutants, but they may lack methodological validation. However, they can be conducted by authorized laboratories, ensuring uniformity and repeatability. Molecular approaches for determining viability in complicated combinations, such as propidium monoazide (PMA)-PCR, are being developed. These, however, are experimental and unstandardized.

In general, fecal recovery is frequently employed as a surrogate sign to indicate adequate dosage for gastrointestinal health goals. Dose range may be achievable for probiotics with easily established effectiveness endpoints (e.g., serum cholesterol levels), albeit the issue of replication in situ may generate interpretation challenges. An IBS trial employing *Bifidobacterium longum subsp. infantis* 35624 examined three dosages, 106, 108, and 1010 CFU was one research that used dose range. This produced an unusual result: the 108 CFU dosage was effective whereas the other two doses were not (Sistek et al., 2006). This unusual finding was most likely caused by the capsules holding the maximum dose did not dissolve, leaving just the results of the other two doses to be evaluated.

An alternative to traditional dosage range research is to review the enormous body of recorded human probiotic trials and compute the doses utilized in each study and the clinical results. A similar investigation was recently conducted, and the results were nuanced. There was evidence of a clear dosage response for some probiotic/health status combinations, but the findings for other combinations were not compelling.

If certain research yields the anticipated health benefit, that dose should be considered the minimal dose for which a health claim should be approved. Products employing a larger dosage should be able to make the same claim, but claims for any dose lower than that evaluated in people should be prohibited. We will not go into greater detail on health claims because it is the domain of regulators. While authorities generally comply with the probiotic definition, they interpret the regulations in their separate jurisdictions differently. An examination of this is outside the scope of this study.

What effect, if any, the distribution format has on the probiotic effect is an intriguing question. Few studies have been undertaken that compares a probiotic administered in multiple matrices to the same clinical objective. Two reviews have addressed the issue, with one indicating that there is no evidence that the delivery matrix has a significant influence on probiotic efficacy and the other suggesting that strain-dependent matrix effects may exist (Aponte *et al.*, 2020).

6.11 FDA PROBIOTICS CLASSIFICATION

The Greek term for probiotics is translated as "for life." The capacity of one organism to promote the development of another was first documented in 1965 when the word was coined for compounds produced by a specific bacteria. Probiotics are currently defined by the Food and Agriculture Organization of the United Nations (FAO) and the World Health Organization (WHO as "living microorganisms that provide a health benefit to the host when fed in appropriate amounts." A bacteria must satisfy four requirements to qualify as a probiotic: it must exist in the human body, survive ingestion, benefit the host, and be safe to consume. Probiotics are most typically found in the GI tract, but they can also be found in the mouth, vagina, urinary tract, skin, and lungs in some situations.

There are various things to consider while categorizing probiotics. Probiotic products are classified as either single-strain or multi-strain. The genus of probiotics is used to define groupings in single strain probiotics. The genus (italicized) and species (in italics) are the two components of a probiotic's scientific name (italicized). The strain's name is occasionally given after the name of the species. The probiotic genus categorization is displayed to the right.

6.11.1 Bifidobacterium

It is a Gram-positive, anaerobic, pleomorphic rod bacterium that does not produce spores or spores. The genus Bifidobacterium is well-known for creating lactic and acetic acids as byproducts of glucose metabolism, which is a common feature (Sidhu *et al.*, 2015). When combined with Lactobacillus species and *Saccharomyces boulardii* probiotic yeast, Bifidobacterium appears to reduce the negative effects of Helicobacter therapy. According to the current investigation's findings, administering *Bifidobacterium infantis* in combination with Lactobacillus acidophilus appears to minimize NEC-associated mortality in severely ill neonates.

6.11.2 Lactobacillus

Lactobacillus is a group of lactic acid-producing Gram-positive rod bacteria that are both obligatory and facultative anaerobes in the human digestive and urinary systems. Lactobacillus is known as Lactobacillus sp. because of its ability to produce lactic acid. Lactobacillus species provide nutritional benefits such as inducing growth factors and increasing mineral bioavailability. Lactobacilli also aid in preserving intestinal integrity by decreasing permeability and maintaining the mucosal barrier.

Bifidobacterium and Lactobacillus are the two most common categories in terms of diversity. Bacillus, Saccharomyces, Lactococcus, and Leuconostoc are not as well-known as Bifidobacterium and Lactobacillus, but their contributions various industries, including fermented foods and dietary supplements, make them an essential component of the probiotics industry.

Conclusion

The probiotics sector includes global research, new product development, marketing, regulatory inspection, consumer interest, and healthcare practitioners' use of probiotics. All parties involved must clearly grasp the standards to guarantee that the term "probiotic" is used correctly. This paper describes the fundamental parameters for a probiotic strain for use in food and nutritional supplements. The same criterion may be used for various probiotic uses. According to approved nomenclatural rules, a strain must be recognized and given a name that can be easily discovered in the scientific literature.

Depending on the probiotic species, several approaches may be applied, and these strategies are likely to change over time. Furthermore, we feel it should be placed in a global cultural collection. The strain must also be safe for the intended use and demonstrate a health benefit based on at least one study that meets scientific standards or, where appropriate, the recommendations and norms of local/national authorities (Arora & Baldi, 2015). Probiotic strains must be present in adequate quantities to provide the stated (and evidence-based) health benefit over the whole shelf life of the product.

The safety, purity, and stability of a product can only be guaranteed if it is manufactured following applicable good manufacturing practices and the product labeling communicates essential information about the product contents to the end user (specific strains, level of live probiotic delivered at the end of shelf life, and statements about health benefits as permitted). It is the product producer's responsibility to make and sell probiotics that comply with local rules and regulations and meet the requirements described above to guarantee that the market does not contain things that abuse the term "probiotic."

CHAPTER 7

BEST PROBIOTIC STRAINS FOR OLDER PERSONS

CONTENTS

There has been a wide range of interpretations of the term "probiotics," derived from Latin and Greek and means "for life" in both languages. A probiotic has been defined as "defined, live microorganisms provided in sufficient quantities to have a beneficial physiological effect on the host," according to recent research. Because of this, probiotics' benefits are becoming more apparent, even though the idea that consuming large numbers of live bacteria can improve one's health and help prevent and treat disease may at first seem odd to speculate (Singh *et al.,* 2011).

Probiotics such as Lactobacilli and Bifidobacteria are two examples. These bacteria produce lactate and short-chain fatty acids such as acetate and butyrate through their metabolic processes. Several probiotic strains, such as *Lactobacillus rhamnosus* GG, *Lactobacillus plantarum* 299v, *Lactobacillus casei Shirota*, and *Lactobacillus johnsonii* La1, are therapeutic in the treatment and/or prevention of intestinal and extraintestinal disorders. Most of these consequences have been thoroughly examined in the most recent studies.

To preserve food (such as sauerkraut), lactic acid-producing bacteria (LAB) have long been used. Most countries have developed their own unique fermented milk styles. This was a completely unintentional practice. As an antidote to aging, Metchnikoff advocated using *Lactobacillus acidophilus* (LAB), which he claimed in the early 20th century was responsible for releasing toxins from putrefactive gut bacteria. In the early 20th century, Metchnikoff's theory was widely accepted. In the interest of public health, this was the first time LAB had been systematically used in a facility (that is, not LAB). People began to pay attention to the concept of functional meals only in the last decade of the twentieth century, which led to an increase in the popularity of probiotics (foods that provide physiological benefits or lower the risk of chronic diseases in addition to their fundamental nutritional value). People's interest waned after a brief burst of excitement (Singh & Natraj, 2021). When it was discovered that germ-free animals lived longer than conventional ones in the 1950s, Metchnikoff's hypothesis that gut flora plays some role in aging was given more credence.

Probiotics are now widely available and frequently purchased by the general public. Bio yogurts and nutritional supplements are the most common active ingredients. Probiotic yogurts and milk accounted for nearly $900 million in sales in Europe in 1997, which is only expected to rise in the future. The concept of prebiotics has emerged only in the last few years that, making it one of the more recent developments in medicine.

Probiotic microbes can thrive with the help of oligosaccharides, which are chemical molecules that act as growth stimulators. As prebiotics cannot be digested by non-probiotic gut bacteria like *Escherichia* and *Bacteroides*, they were chosen as the best candidates. There are two types of prebiotics: FOS and GOS, which are fructose- or galactose-based oligosaccharides found in breast milk and some plants (Rusu *et al*., 2019). It is possible to obtain prebiotics either chemically or biologically (such as Jerusalem artichokes and onions). In addition to probiotics, they can also be added to meals on their own.

For the elderly, this study aims to highlight the benefits of probiotics and prebiotics in constipation, malnutrition, and a compromised immune system, among other things.

This semisolid texture and robust flavor are due to the precipitation and denaturant of casein by lactic acid. It is common to add sugars and flavorings to tamp down this flavor. In the fermentation process, "bio yogurts" use other probiotic bacteria, most commonly *Lactobacillus acidophilus*, which is very similar to those used in the production of traditional yogurt.

Many beverages, including fermented milk and fruit juices with added vitamins, contain probiotic bacteria. Yogurt and probiotic beverages are available in various flavors at supermarkets and other retail outlets.

If you are looking for another source of probiotics besides yogurt, you can find them in freeze-dried bacteria in dietary supplements. Many drugstores and health food stores carry these products and online retailers like Amazon. Consider that many probiotic products contain bacteria that cannot be identified or whose quantity varies, and that most of these types do not have a documented history of probiotic use. Consider this. Only a few of the hundreds of lactobacilli strains have been scientifically proven to have health benefits, but they are still widely available (Reid, 2016).

7.1 PREBIOTICS

FOS and GOS, commonly used as prebiotics in probiotic formulations, do not provide any health benefits at the current doses. Although commercially available, synthetic prebiotic oligosaccharide compounds have yet to be incorporated into consumed foods in the UK. Artichokes, onions, and chicory are some of the best sources of prebiotics right now (Qi *et al*., 2019).

There is far too much malnutrition in the world. 30 %of the world's population is estimated to have this ailment, and it is not confined to developing countries. Hospitalized patients across Europe and the United States were found to be underweight or considered dangerous by two studies conducted over the past decade.

The plight of the elderly is well-known. Nearly 7 %of people who appear healthy and are still living at home fall into this category, which rises with age. Another risk factor is being sick or confined to one's home. It is thought that malnutrition damages the gut epithelium, reducing gut-mediated immunity, nutrient absorption, and appetite. Bacteria from the intestines can enter the bloodstream, increasing the risk of septicemia and other infections.

The World Health Organization recommends adding probiotics to yogurt to improve the nutritional status of children. *Lactobacillus acidophilus, Lactobacillus delbrueckii var bulgaricus*, *Streptococcus salivarius var thermophilus*, *Lactobacillus Plantarum* 299v, and *Bifidobacterium lactis* Bb12 are among the 15 effective probiotic species. It would be exciting and informative to investigate whether these probiotic species can help alleviate malnutrition in the elderly. Lactose intolerance is prevalent, and age tends to increase the symptoms. Avoiding milk may prevent these in general, but doing so may significantly restrict the consumption of vital minerals like calcium.

Low levels of -galactosidase (lactase) in the intestinal mucosal cells cause the sickness. It has been found that fermented milk products, such as yogurt, are better tolerated than raw milk, and it has been suggested that this is at least partially owing to the -galactosidase activities of the bacteria employed to ferment the yoghurt (*L. delbrueckii var bulgaricus* and *S. salivarius* var *thermophilus*) (*L. delbrueckii* var *bulgaricus* and *S. salivarius* var *thermophilus*) (*L. delbrueckii var bulgaricus* and *S. salivarius* var *thermophilus*). After these organisms have gone through the stomach and been killed by pancreatic secretions, an enzyme is created (Prins, 2011). Even though the well-known probiotic strain *L. rhamnosus* GG cannot digest lactose, the bacteria linked with bio yogurts (*L. acidophilus* and *B. bifidum*) are bile stable and hence less helpful in curing lactose intolerance. Therefore, live yogurt is indicated when treating elderly individuals with lactose intolerance.

Szilagyi draws attention to the parallels between the symptoms of irritable bowel syndrome and lactose maldigestion and warns that either condition may be misdiagnosed. Therefore, the revelation that probiotics can help heal irritable bowel syndrome is promising.

According to animal and human research, prebiotics such as lactulose, inulin, FOS, and GOS can improve calcium bioavailability when consumed in large amounts (10-40 g per day) (10-per 40 g per day) (10-40 g per day). In a group of postmenopausal women, Van den Heuvel et al. observed that GOS at a dosage of 20 g/day improved calcium absorption by 16 %. It is fair to presume that the extra calcium was retained because there was no increase in calcium excretion in the urine. These trials show hope that long-term prebiotic usage can prevent or cure osteoporosis, but additional research is necessary.

In addition to the disorders described above, scientific studies have proved the usefulness of probiotics for a large diversity of conditions. Even though older adults were not particularly included in this study, it is feasible to presume that the results also apply to this age range.

There is a good likelihood that LAB can prevent cancer. Animal and *in vitro* studies show that changing digestive enzymes to limit the formation of possible carcinogens such as nitrite and azo compounds may have anticarcinogenic advantages. As noted in the earlier section, the potential of probiotics to improve the activity of NK cells, one of whose targets in tumor cells, may also contribute to the prevention of cancer growth (Pratap et al., 2020). Epidemiological studies, some of which imply a decrease in the incidence of certain kinds of cancer related to the frequent consumption of fermented milk products, give grounds for optimism. According to two randomized clinical studies done in Japan, *L. casei Shirota* substantially decreased the recurrence risk of bladder cancer in males.

7.1.1 Lactobacillus reuteri

Lactobacillus reuteri was named in honor of Gerhard Reuter, a German scientist who dedicated the majority of his career to studying the human gut microbiota and specifically Lactobacilli and Bifidobacteria. His team was confronted with a challenge that we face today in microbial ecology: how can we get to the small intestine with the least amount of damage possible? Lactobacilli were found in the small intestines, and feces of many of the subjects studied, including children and adults, using controlled capsule sampling methods and culture from patients who died suddenly. There was a significant population of Lactocci in both the stomach and duodenum, two regions of the digestive system previously thought to be free of microbes for an extended period (Pandey *et al*., 2015). Although the presence of bacterial communities in these regions is now widely accepted, this assertion was initially met with skepticism fifty years ago.

According to Reuter's findings from human experiments, the gut microbiome is home to several species of Lactobacillus, including *L. reuteri* and *Lactobacillus gasseri*. Aside from jejunum/ileum and faeces, *L. reuteri* strains were rarely found in any other part of the gastrointestinal system in human patients. Because *L. reuteri* was isolated from the same individuals repeatedly, it is clear that *L. reuteri* is a stable component of the human microbiome. The following section will discuss ecological niche occupied by *L. reuteri* strains in humans.

Lactobacillus reuteri, one of the most common probiotics, may produce antimicrobial compounds. Bacterial growth, adhesion, and invasion are all halted by these compounds. Increasing intestinal epithelial cells' survival and proliferation may help maintain the barrier's integrity. TNF-alpha and other proinflammatory cytokines may be inhibited, and protective cytokines such as IL-10 and TGF- may be increased in their place. Finally, the expression of micro opioid and cannabinoid receptors can be influenced by certain probiotics, which have an analgesic effect on stomach discomfort.

Endotoxemia, which is linked to liver damage caused by alcohol consumption, can be reduced in cirrhosis while liver function can be improved. These people received *Lactobacillus plantarum* and *Bifidobacterium bifidum*. The probiotic strain *Lactobacillus casei Shirota* was found to reverse neutrophil dysfunction in patients with compensated liver cirrhosis. *Lactobacillus subtilis* and *Streptococcus faecium* were also found to reduce the amount of LPS in the guts of patients suffering from alcoholic hepatitis. *Lactobacillus rhamnosus* and *Lactobacillus acidophilus* reduced TLR4 levels in four-week mouse studies compared to the control group (Parvez et al., 2006). TLR4 is, as previously stated, a critical link in the inflammatory cascade that sets off the disease-inducing process characteristic of alcoholic liver disease. Proinflammatory cytokines such as IL-1B and TNF-alpha were also found to be reduced in the study.

An indole-3-aldehyde metabolism by *Lactobacillus reuteri*, activates the aryl hydrocarbon receptor (Ahr) on ILC and causes the production of the barrier-protecting cytokine IL-22, results in the ability of Lactobacillus to break down tryptophan. It is obvious that the Ahr is essential to provide a docking site for ligands generated by microbiota and responsible for activating the immune system. Studies using probiotic strains of bacteria, such as *E. coli* Nissle, have demonstrated its capacity to alter the direction of immune responses, moving them from Th2 to Th1 protective immunity. In the past ten years, it has come to light that intestinal bacteria are the source of thymic stromal lymphopoietin (TSLP), a protein that plays a critical role in the development of FOXP3 T-regulatory cells and is essential for the regulation of inappropriate Th17 responses. Recently, we were able to prove that *L. reuteri* is effective in preventing age-related sarcopenia as well as cancer-induced cachexia. The beneficial effects on skeletal muscle coexisted with higher growth hormone levels, decreased systemic inflammatory tone, and improved forkhead box N1 (FoxN1) expression in the thymus gland.

FoxN1-deficient animals (athymic nude) cannot avoid sarcopenia in response to *L. reuteri* feeding, which suggests a FoxN1-mediated mechanism. There is also a trophic effect that oxytocin performs in skeletal muscle. Experiments with mice have shown a natural decline in plasma levels of oxytocin associated with aging and that the hormone is essential for the regeneration of muscle tissue (Novik & Savich, 2020). Animals with a deficiency in oxytocin develop sarcopenia at an earlier age, whereas treatment with an oxytocin antagonist inhibits the regeneration of muscle tissue.

In contrast, the systemic injection of oxytocin improves the MAPK/ERK signaling pathway, stimulating old muscle stem cell activation and proliferation and muscle regeneration. Several studies on mice have demonstrated that *L. reuteri* increases bone density (Collins *et al.*, 2016; Mccabe, Irwin, Schaefer, & Britton, 2013) and prevents bone loss in both menopausal ovariectomized and type 1 diabetic animal models. In mice that have had their ovaries surgically removed, the effects of *L. reuteri* are mediated via a reduction in the osteoclastogenesis caused by effector CD4+ T cells. *L. reuteri* prevented the loss of osteoblast quantity and function caused by TNF- in diabetic male mice.

Oxytocin possesses powerful bone-building and anabolic properties. Oxytocin is essential for maintaining a healthy equilibrium between osteoblast and osteoclast activity. It promotes bone formation by stimulating osteoblast activity. Even though it promotes the production of osteoclasts, it prevents the osteoclasts from engaging in bone resorption.

7.2 BEYOND THE IMMUNE SYSTEM, GENETIC MODIFICATION AND BONE HEALTH

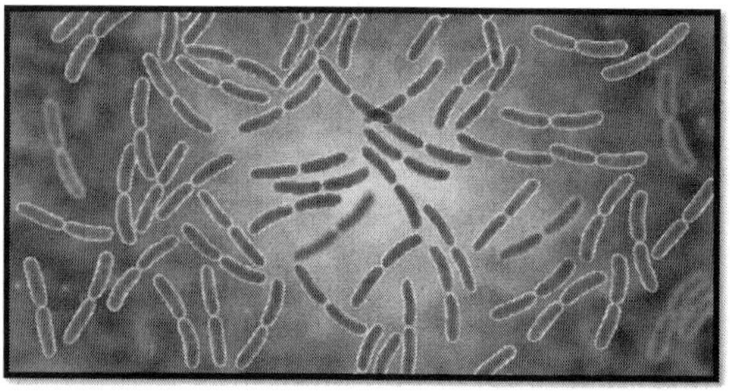

Figure 7.4: *Lactobacillus reuteri* observed under a microscope.

The make-up of genetically modified organisms and how they are regulated may affect bone health by altering calcium absorption and the production of serotonin generated from the gut. In individuals with healthy bones and teeth, *Lactobacillus reuteri* increases the quantity of 25-hydroxyvitamin D in the blood, which regulates calcium absorption and promotes bone health. It is not quite obvious how this probiotic affects the vitamin D level in the body, but it does. This could be because of a change in the environment of the gut, which favors the absorption of vitamin D, or it could be because of an indirect impact on improved hepatic 25-hydroxylase activity or 7-dehydrocholesterol concentration, which results in reduced absorption of cholesterol from the diet and bile (Nichols, 2007).

It has been hypothesized that lower levels of vitamin D consumption are associated with altered GM profiles. If this is true, the connection between genetically modified organisms and vitamin D might have the opposite effect. One such possible way in which GM improves bone health is by increasing the amount of calcium that is absorbed by the body. It is general knowledge that maintaining a healthy calcium balance is one of the most important factors in achieving a satisfactory peak bone mass, which, in turn, protects against the development of osteoporosis in later years. Calcium absorption is improved when fibers are consumed through the diet. After being fermented by GM, fibers stimulate calcium absorption in two ways: first, they lower the pH of the gut, which in turn restricts the production of calcium phosphates and raises calcium absorption; second, they increase the production of short-chain fatty acids (SCFAs) like butyrate. The impact of SCFAs may be more nuanced than the effect they have on the pH of the gut, and they may be able to boost calcium transport through changes in signaling pathway activity.

One such proposed mechanism through which genetic modification (GM) contributes to improved bone health is the effect it has on the generation of serotonin (5HT) in the gut. It has only been discovered very recently that serotonin has a dual effect in the regulation of bone mass, and this action differs depending on the region where this molecule is produced. The gut-derived 5HT (g5HT), which GM changes, has been ascribed to the function of acting as a regulator of bone mass. In the duodenum, enterochromaffin cells are responsible for the synthesis of g5HT. This process is primarily controlled by GM since SCFAs promote the production of g5HT. 5HT has been shown to influence bone cells and, in particular, a suppressive effect on the production of osteoblasts (Attri et al., 2021). These observations, show that GM's capacity to modulate g5HT may offer a feasible treatment strategy for enhancing bone health. A pharmacological reduction in g5HT production led to the prevention of osteoporosis, which was mediated by improved bone formation. This was the case with ovariectomy-induced bone loss. However, there is a significant debate regarding the facts on the impact that 5HT has on bone health.

7.3: LACTOBACILLUS RHAMNOSUS

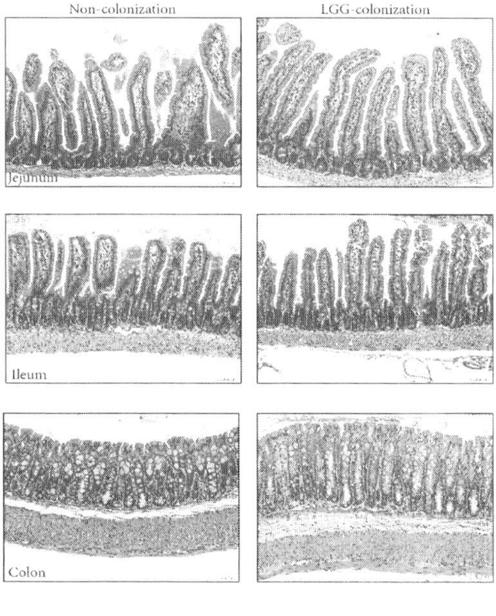

Figure 7.5: Lactobacillus rhamnosus.

GBS is an opportunistic bacterium widely detected in the vaginal canal and is a primary factor in preterm birth and illness in neonates. It is hypothesized that the presence of commensal Lactobacillus species in the vaginal canal works as a line of defense against infections such as GBS. Some research has studied whether or not lactobacilli can impact the pathogenicity of GBS and, if so, how. Using an *in vitro* model system, this experiment aimed to evaluate whether or not *Lactobacillus rhamnosus* exhibits a possible protective role against GBS. According to the results, GBS is more successful at separating HeLa cells than in situations when *L. rhamnosus* was present. As observed in the SEM photos, GBS induces a decrease in both the length and the number of microvilli that are seen on the surface of HeLa cells (Narang et al., 2011). It also causes a decrease in the size of the vesicles that are generated. *L. rhamnosus* can inhibit the breakdown of microvilli and vesicles that GBS induces. Additionally, GBS can degrade F-actin fibers in HeLa cells even when *L. rhamnosus* is absent. These data imply that GBS infection affects the host cell's cytoskeleton, which leads to the conclusion that *L. rhamnosus* has a role in inhibiting GBS colonization of the host.

The word "inflammation" refers to a state in which chronic, low-grade inflammation culminates in the development of age-related disorders. Studies have indicated that taking probiotic supplements throughout the aging process can help minimize the likelihood of getting ailments that are connected with increasing age. However, there is considerable controversy surrounding whether or not providing probiotics to newborns, and early children can help adults live longer. The impact of *L. rhamnosus* GG (LGG) colonization on inflammatory disorders in children was examined. In this study, pregnant mice were administered either LGG live bacteria (LC group) or LGG fixed bacteria (LC group) beginning on the 18th day following conception and continuing through birth (NC group). For the first five days of their lives, newborn mice were administered 107 CFU of either live or fixed LGG.

Colonization with LGG has been discovered in the faeces of newborns who were just 3 weeks old. The 16S rRNA sequencing investigation that was done on neonates at 3 weeks of age indicated that early LGG colonization might have an impact on the makeup of the gut microbiota as well as its diversity. Surprisingly, the positive effects of LGG colonization on the microbiota continued up to 8 months in the research individuals (Nami *et al.,* 2015). The number of bacteria associated with improved lifespan (Lactobacillus, Bifidobacterium, and *Akkermansia muciniphila*) considerably increased in the LGG colonization group. Furthermore, LGG colonization increased both the number of bacteria that make short-chain fatty acids (SCFAs) and the synthesis of SCFAs in the cecum.

In mice pups aged 8 months, LGG colonization preserved the intestinal barrier, increased antioxidant defense, reduced epithelial cell DNA damage, and lowered low-grade intestinal inflammation. In a mechanical sense, LGG was able to upregulate the Sirtuin1 (SIRT1)/Adenosine 5-monophosphate-activated protein kinase (AMPK)/Peroxisome proliferator-activated receptor coactivator 1- (PGC-1) pathway and repress nuclear factor-kappa B (NF-B) activation; however, silencing the SIRT1 gene decreased the protective effect of LGG. The first paragraph of the introductory. The aging process is complex, defined by a steady reduction in physiological function and adaptive ability (Ayichew *et al.*, 2017). This increases the risk of acquiring several chronic illnesses, including cancer. It is a commonly held concept that good aging and longevity result from complex interactions between genetic and nongenetic (e.g., lifestyle, sociodemographics, and living circumstances) variables.

Both of these types of factors are intimately connected with gut flora. The great majority of microorganisms that resided in the nutrient-rich environment of a human gut were bacteria that did not harm human health but benefited it. Shortly after birth, the population of gut bacteria comparable to those seen in adults is created, and it continues to be generally constant throughout maturity. The first few years of a person's existence are crucial for the preservation of their health as well as the development of their metabolism. The microbiota of mothers can be passed on to their offspring through the mother's feces, through the birth canal, through the skin, or by nursing. Therefore, the microbial makeup of babies is greatly impacted by the microbiota present in the mother's gut (McKenzie *et al.*, 2018). On the other hand, the stability of the gut microbiota may be affected by variables associated with aging, such as changes in physiology, lifestyle, and health status.

The elderly have a lesser microbial diversity than healthy adults, and the make-up of each individual's microbiota differs greatly. The relevance of microbial ecology in influencing health condition and lifespan is brought into focus by these findings. By modifying the makeup of gut microbiota and systemic immunity, probiotics such as Bifidobacterium and Lactobacillus have been proven to prolong life expectancy and diminish age-related diseases. It has been proven that oral treatment of . *plantarum* HY7714 or

B. breve B-3 helps reduce skin photoaging. *L. brevis* OW38 decreased gut microbiota lipopolysaccharide (LPS) synthesis, NF-B activation, and p16 expression in old mice, which may attenuate colitis and cognitive deficits associated with aging.

Supplementation with *L. plantarum* WCFS1 decreased the age-related reduction in the colon mucus barrier. LGG is a Gram-positive bacteria isolated from a healthy human digestive tract demonstrating robust intestinal cell adhesion. LGG has a long history of usage as a dietary supplement for pregnant women and newborns. Clinical investigations have shown that perinatal LGG treatment for mothers helps avoid eczema in high-risk children and that feeding formulas containing LGG to term infants may be superior in encouraging normal development. LGG stimulates anti-inflammatory pathways of resident microorganisms, while the preventive advantages of LGG in early life on aging-related illnesses have not been fully investigated (Bowe et al., 2014). Recent research reveals that LGG colonization in newborn mice improves health in the early stages of aging, suggesting that LGG colonization in infancy might be employed as a prophylactic approach against age-related disorders. Therapie utilizing Mice and LGG In a broth of Lactobacillus MRS, live LGG was cultivated. LGG preserved with glutaraldehyde was produced in line with prior studies. Twenty (8-week-old) mature C57BL/6 female mice were procured from Beijing HFK Bioscience Co., Ltd. and kept in five pathogen-free (SPF).

The animal rooms received HEPA-filtered air at a rate of 15 air changes per hour, and the temperature was maintained between 20 and 26 °C. Mice were given a regular mouse diet and maintained on a natural light-dark cycle. All mice drank ad libitum regular, sterile water, and the water was replenished weekly or as needed. After mating with C57BL/6 males for two weeks, 18 pregnant mice with the same conception time were randomly split into two groups and gavaged with either 108 CFU of live or fixed LGG from day 18 of gestation until natural delivery. All subsequent offspring were born vaginally, and 18 puppies were collected in the NC group and 20 in the LC group.

To prevent gender differences, 14 kids were randomly recruited for each group. The young mice were injected 107 CFU of either live or fixed LGG for five days beginning the day after birth. In the third week, seven mice (3 males and 4 females) were chosen randomly from each group and slaughtered; the other animals were slain eight months later. The Animal Ethical and Welfare Committee of Tianjin Medical University investigated and authorized all animal research. LGG Colonization Detection: To determine LGG colonization, the feces of 3-week-old offspring were collected. Genomic DNA was obtained from growing bacteria using the DNA Rapid Extraction Kit and then used for polymerase chain reaction (PCR) amplifying the 16S rRNA gene with LGG-specific primers. The migration profile of PCR-amplified products was examined using denaturing gradient gel electrophoresis.

7.4 BIFIDOBACTERIUM LONGUM SSP. LONGUM

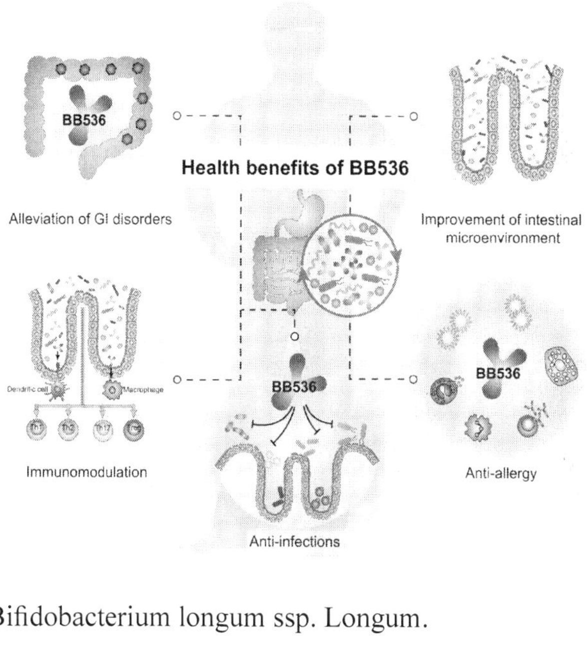

Figure 7.1: Bifidobacterium longum ssp. Longum.

Bifidobacteria are one of the key components of human microbiota that are assumed to have a function in the upkeep of human health in some way. The colonization and cell quantity of different species of Bifidobacterium that dwell in the human gut could alter as a person matures.

However, the reasons for the sequential changes in Bifidobacterium species that may be detected in humans ranging in age from newborns to centenarians are still a mystery. We explored the changes in the composition of the Bifidobacterium species in the gut that occur during a large age range (Mahasneh & Abbas, 2010). The analysis of fecal samples from 441 healthy Japanese individuals ranging in age from 0 to 104 years was performed with real-time PCR utilizing primers specific to the species.

The *B. longum* group had the greatest detection rate, and its presence was identified in humans of all ages, from babies to centenarians. About 70 %of children under the age of three were discovered to have *B. breve* in their bodies. Following weaning, the *B. adolescentis* and *B. catenulatum* groups predominated in the population. The presence of *B. bifidum* has been documented at virtually every age.

Compared to other age groups, the elderly demonstrated a greater rate of *B. dentium* detection. It was observed that 11.4 %of the volunteers, whose ages were restricted, harbored the *B. animalis sp. lactis* bacterium. Only nine persons tested positive for the *B. gallinarum* group, whereas neither B. minimum nor B. Mongolians could be discovered in any age group examined.

Researchers showed that the presence of specific Bifidobacterium groups was connected to the presence of many other Bifidobacterium species and subspecies.

There were inter-species links between all of the species, except the *B. animalis* sp. *lactis* population. These studies highlighted patterns and transition points in the compositional changes of Bifidobacterium species that occur with age, and the findings imply that some of these species may have symbiotic relationships in the gut microbiota (Li *et al*., 2016).

It is commonly acknowledged that representatives of the genus Bifidobacterium may be found in the human gastrointestinal (GI) tract, where they are assumed to play critical functions in sustaining human health.

Bifidobacterium is the genus that is identified in the largest numbers in the gut microbiota of breastfed newborns. However, the population of this species shows a steep drop after weaning and continues to do so with age. Additionally, it was observed that the species-level composition of Bifidobacteria varied across the life phases of an organism. About ten distinct species and subspecies of Bifidobacterium have been detected residing in human stomachs. According to research undertaken in the past, *B. breve* and *B. longum* are the main species that may be discovered in the gut microbiota of neonates. It has been established beyond a reasonable doubt by Avershina et al. that a shift from a microbiological profile associated with childhood to one associated with maturity takes place between the ages of one and two years. In addition, they argued that the keystone of *B. breve* strains was to cause of this alteration.

There was a negative link between *B. breve* and *B. longum* in the adult population. *B. breve* was found significantly abundant in the gut microbiota community present in neonates throughout the first year of life.

The adult gut microbiota is dominated by the *B. adolescentis* and *B. catenulatum* groups and the infant-associated *B. longum* ssp. *longum*. This is in contrast to the newborn gut microbiota, which is dominated by the *B. longum* ssp. *longum* group. These adult-type dominating Bifidobacterium species continued until later in life, with the preponderance of *B. dentium* species increasing. Recent research done by Wang et al. has shown that the species of intestinal Bifidobacterium that were detected in centenarians of Chinese descent were unique from those that were discovered in young seniors.

Despite common faecal Bifidobacterium species in younger elderlies and centenarians, such as *B. dentium*, and *B. longum* ssp. *longum*, *B. thermophilum*, members of the *B. catenulatum* group, and *B. adolescentis*, centenarians were found to have certain species only found in centenarians, such as *B. minimum, B. gallinarum/ B. pull*

Although some research has compared the Bifidobacterium species identified in healthy adults to those detected in newborns and/or the elderly, no studies have explored the sequential changes that occur in Bifidobacterium species from the time of birth to the age of 100 years or more. The purpose of this work was to obtain a greater grasp of the alterations in the makeup of Bifidobacterium species when they approach adulthood (Lin *et al*., 2019). We studied the age-related sequential changes in the species composition of Bifidobacterium in 441 healthy Japanese volunteers ranging in age from zero to one hundred and four.

The exploration of the human gut microbiota and how it impacts human health is a prominent topic of research in the discipline of microbiology.

Numerous research, beginning with the earliest investigations utilizing culture-dependent methodologies and proceeding to the present omics approach, have been carried out to evaluate the make-up and functionality of the gut microbiota.

Several researchers have conducted these investigations. Numerous research has centered on the complex microbiota and the functions that it plays in the treatment and prevention of illness.

According to Biavatti and Mattarelli, members of the genus Bifidobacterium are present in the human host practically all of the time and serve a significant job in the gut from birth onward.

This genus has gained much interest because Bifidobacteria have historically been regarded as probiotics.

Differences in species composition, diversity, and relative abundance have been explored in connection to a range of life phases and disorders.

This mini-goal review aims to emphasize the compositional changes of bifidobacteria as they relate to different life phases, highlighting both their positive advantages and the probable protective implications of their presence against various illnesses situations (Kumar, 2013).

The review will also underline the relevance of bifidobacteria's putative preventive implications against several illness conditions.

Microbiota that dwells in the digestive tract. According to Gurner and Malagelada (2003), gut microbiota substantially influences on human health.

It has been related to trophic, protective, and metabolic activities for a very long period, in addition to, more recently, the gut-brain axis and the liver-gut axis. We now have fully understand the microbiota that lives in the gut as a direct result of recent breakthroughs in sequencing technology and the proliferation of other "omics"-related scientific domains.

It was traditionally assumed that the creation of the gut microbiota took occurred at birth; however, identifying bacteria in the placenta or amniotic fluid (Collado et al., 2016) implies that colonization occurs during the development of the fetus.

In addition, the process is governed by various different variables, all of which might upset the equilibrium of the microorganisms. Early life is defined by the prevalence of the phyla Actinobacteria, Proteobacteria, and Firmicutes, which are noted for their lack of variety and complexity. Between the ages of one and two, the most dramatic changes occur in the makeup of the microbiota that dwells in a child's gut (Voreades *et al.,* 2014).

It is estimated that by the third year of a person's life, the gut microbiota will have formed and consist of 1014 species.

These species will comprise bacteria, eukarya, viruses, and archaeal members. Between 80 and 90 % of the gut microbiota at the phylum level may be attributable to the Firmicutes and Bacteroidetes.

At this point, the variety of species and strains is unusually large, and there is a substantial degree of variance across people (HMP, 2012). On the other hand, the metabolic process and activity of the gut microbiota are surprisingly conserved (Qin *et al.,* 2010).

According to Imahori (1992), the process of becoming older may be characterized as "the regression of physiological function paired by advancement in age." This normal process can result in alterations to the GI tract, immunosenescence, and, in extremely rare circumstances, malnutrition. In addition, adjustments to one's food, lifestyle, and usage of medications have an inescapable influence on the make-up of the geriatric microbiota and the roles they perform. Even though the variety of microorganisms declines with age, the phyla Bacteroidetes and Firmicutes continue to dominate. In contrast, the relative abundance of numerous other phyla, most notably Proteobacteria, has grown. Bifidobacteria levels are lower in adulthood, ranging from 2-14 %relative abundance while maintaining stability. Using quantitative polymerase chain reaction (q-PCR), Gueimonde et al. (2007) revealed that newborns had much greater quantities of Bifidobacterium than adults.

Their idea, which is corroborated by previous research findings, suggests that *B. longum* is the species that appear most frequently. Nonetheless, Matsuki et al. (2004) discovered that *B. adolescentis* and *B. catenulatum* were more frequent in the adult population than in the juvenile population.

According to Chaplin et al. (2015), the rate at which *B. adolescentis* was isolated grew with age, but the rates at which *B. bifidum* and *B. breve* were isolated decreased with age. Because of high inter-individual variability, variations in food and lifestyle, and the vagueness of the word "elderly," there is no consensus on what defines an old-age-specific gut microbial composition.

Nevertheless, certain trends continue, such as the decrease in Bifidobacteria in the elderly population, which has been supported by many studies employing various methods of scientific investigation. According to Woodmansey et al. (2004), as people became older, there was a decline in both the range of species and the quantity of Bifidobacteria in their stomachs.

This drop, which has been associated with a reduced amount of adhesion to the intestinal mucosa, may be attributed to variations in the microbiota or the structure of the mucus. However, it is not definite (He *et al.*, 2001). In addition, a study has documented the makeup of Bifidobacteria in community centenarians; nevertheless, the conclusions are still somewhat controversial. Microbiota compositions of centenarians in a European community were comparable to those of adults (Biagi *et al.*, 2010); nevertheless, centenarians revealed larger proportions of Bifidobacteria than younger elderlies from a Chinese region. According to the species, the most prevalent among Italian centenarians were *B. longum, B. adolescentis*, and *B. bifidum* (Drago *et al.*, 2012), but in Chinese centenarians, *B. dentium* was the most common. The makeup of Bifidobacteria can also be modified by other external factors that are indirectly associated with aging.

Long-term use of antibiotics in the elderly population has been found to have a major influence on the makeup of the gut microbiota, resulting in a drop in the number of Bifidobacteria. This Bifidobacteria drop has been associated with an increased risk of acquiring urinary tract infections.

Even though antibiotics continue to play an important role medicine, one treatment option that has been investigated is the administration of probiotics to restore a healthy balance between the Bifidobacteria population and the microbiota in the gut that has been altered as a result of antibiotic treatment. According to Van Tongeren et al. (2005), there was no change in the amounts of Bifidobacteria observed in older persons with varying degrees of frailty.

Bartosch et al. (2004) observed that older persons living in the neighborhood and those hospitalized have a propensity to act comparably.

7.5 BIFIDOBACTERIUM BIFIDUM

The bacterium known as *B. bifidum* is a Gram-positive, anaerobic organism that neither forms spores nor moves about. The bacteria that adopt the shape of rods can exist on their own, in pairs, or even in groups. It is possible to locate the bulk of the *B. bifidum* population in the colon, the lower small intestine, breast milk, and very occasionally, in the vagina. The human gastrointestinal system is home to the helpful bacteria *B. bifidum*. It is a sign of bad health when levels in the human stomach are low or when it is missing. Oral ingestion of *B. bifidum* can result in alterations to the microbiota of the gut. Oral *B. bifidum* is also utilized in treating hepatic and intestinal illnesses, as well as in the stimulation of the immune system and preventing some tumors (Kukkupuni *et al.*, 2015). As people become older, their levels of *B. bifidum* decline. When *B. bifidum* levels diminish, other bacteria such as Lactobacilli, Enterococci, and Enterobacteria begin to thrive.

All of these factors contribute to an increased risk of developing cancer in older people and a reduction in the capacity of their livers to function optimally and efficiently.

The Bifidobacterium microbiome of an older adult's gut may differ from that of an adult, increasing the older adult's risk of infections and gut barrier failure.

Certain probiotics can strengthen the barrier that protects the intestinal tract from outside substances. During this randomized, placebo-controlled intervention study, 66 senior citizens were given either a fermented oat beverage that contained the probiotic strains *B. longum* 46 and *B. longum* 2C or a non-fermented beverage that served as a placebo. The participants consumed both beverages for six months (Kim & Park, 2017).

The feces samples were taken before, during, and immediately following the intervention. Plate counting and species-specific q-PCR were the methods that were utilized to determine the levels of faecal Bifidobacteria. Bifidobacterium levels were relatively high in elderly patients, and a wide variety of species was discovered. The administration of probiotics resulted in a significant increase in the levels of Bifidobacteria.

The levels of *B. catenulatum, B. bifidum,* and *B. breve* were all significantly higher than usual.

Alterations in the microbiota composition were associated with the consumption of the fermented oat beverage.

Last but not least, probiotic treatment has the potential to alter the Bifidobacterium microbiome of older adults.

Bifidobacterium bacteria may be found in certain healthy senior populations in greater abundance and diversity than was previously thought.

Maintaining good health throughout a person's lifetime requires careful attention to the microbiota in the gut.

Microbiota diversity is important for disease resistance because it fosters interaction between the host's microbiome and immune system.

It is essential for the host's health that commensal microorganisms affect colonization, disease resistance, and healthy diversity (Khoruts, 2018).

In particular, gut Bifidobacteria have garnered interest due to their connection to the concept of "typical healthy intestinal microbiota" and their potential health benefits. Changes in the composition of the gut microbiota that are caused by aging include decreased species diversity and decreased numbers of beneficial bacteria.

The elderly may have altered gut flora due to changes in nutrition, alterations in the immune system, hospitalization, prolonged intestinal transit times, a lack of physical activity, repeated infections, and frequent use of antibiotics.

It has been demonstrated through conventional culture methods that older people have significantly lower levels of the species Bifidobacterium than healthy adults.

Investigations that used methods that were not dependent on culture, on the other hand, did not find any reduction in the number of Bifidobacteria.

It is common for older adults, particularly those who live in nursing homes or other types of residential care facilities, to experience both constipation and diarrhea (Jugan *et al.,* 2017).

As a consequence of this, it is desirable to adjust and balance the microbial composition as well as the gut function of the elderly.

It has been hypothesized that probiotics could act as function modulators on the gut microbiota.

Although probiotics may benefit this population, few studies have been conducted in this age range.

We anticipated that the specific probiotics chosen for this target group would improve the gut barrier by increasing the composition and diversity of the Bifidobacterium microbiota (Johnston *et al*., 2012).

An oat product fermented with two probiotic Bifidobacterium strains specifically designed for the elderly was evaluated using a species-specific qPCR method and a traditional culturing method to improve the gastrointestinal Bifidobacterium composition of elderly nursing-home residents. Probiotic Bifidobacterium strains were used in the fermentation process.

CHAPTER 8

COMMON SPECIES OF PROBIOTICS ON FOOD AND SUPPLEMENT LABELS

CONTENTS

Probiotics can be added to a healthy and balanced diet through naturally occurring foods such as yogurt and fermented beverages. The diet can be supplemented with dietary supplements, which come in pill, capsule, liquid, and powder form. Probiotic supplements are not all the same. Depending on the strain, different strains of the same species of probiotics may have different efficacy and safety profiles. On the other hand, probiotic regulations in the United States are still in their infancy (Hill, 2010). This review focuses on probiotics and examines the rules, issues, and potential future directions.

Understanding the significance of probiotic bacteria and the mechanisms by which they exert their beneficial effects has become increasingly important in recent years. Ilya Mechinikov, a Russian microbiologist and Nobel Prize winner, claimed in the early 1900s that Bulgarians' high consumption of fermented probiotic foods contributed to their long life and good health. He concluded because the ratio of "bad" to "good" bacteria in the human gut determines overall health. He identified and discovered the beneficial bacteria used in food fermentation. According to French pediatrician Henry Tissier, healthy children had a high number of these "bifid" bacteria in their stool, whereas children with diarrhea had a low number of these unusual Y-shaped bacteria. This study also discovered that bifidobacteria dominated the gut flora of both breastfed babies and adults (Harzallah & Belhadj, 2013). *Bacillus bifidus* was initially known as Bifidobacterium, but this was changed later. He proposed giving these bacteria to diarrhea patients to restore healthy gut flora. This effect was attributed to Bifidobacteria displacing disease-causing bacteria that produce proteolytic enzymes. Metchnikoff and Tissier proposed probiotics in their scientific works.

A German professor named Alfred Nissle isolated a non-pathogenic strain of *Escherichia coli* from the feces of soldiers who did not develop enterocolitis as part of the investigation into the 1917 outbreak of shigellosis in Germany. In treating intestinal disorders, viable non-pathogenic bacteria were frequently used to alter or replace the intestinal microbiota. One of the few examples is E. coli strain Nissle 1917, a non-laboratory probiotic. The term "probiotic" was first used in the 1960s to describe substances produced by microorganisms that promote the growth of other microorganisms. To emphasize the microbiological nature of probiotics, Fuller redefined the term "live microbial feed supplement that improves the intestinal balance of the host animal in 1989, describing viable bacteria cultures that, when applied to an animal or a human, improve the host's natural flora (Charalampopoulos & Rastall, 2009). When consumed in sufficient quantities, "living microorganisms can confer health benefits on the host beyond inherent basic nutrition," according to a more recent but unconfirmed definition.

Intestinal bacteria, as well as bacteria isolated from fermented dairy products, have been included in the definition of "probiotics." Probiotic bacteria are now found in a wide variety of foods, supplements, and medications. Because LAB and fermented milk have been around for a long time, they are almost entirely found in dairy products such as fluid milk and yogurt. Lactobacillus and Bifidobacterium are abundant in these products. Recently, probiotic bacteria have been added to non-dairy foods such as chocolate, cereals, beverages, fruits, and vegetables. As probiotic product development continues in academic and industrial centers worldwide, non-traditional food products containing probiotic bacteria are expected to enter the market soon.

8.1 PROBIOTIC BACTERIA DEFENDERS

Figure 8.1: Probiotic label.

Probiotic bacteria are commonly sold as food or feed supplements, so they must be safe. Over many years of safe use as food starter cultures, the safety of these microbes has been established. Bacteria such as *Lactobacillus*, *Leuconostoc*, and *Pediococcus* species are used in food processing and human consumption. GRAS (generally recognized as safe) LAB has long been used in food processing, and there have been no reports of adverse effects from consuming these bacteria. LAB are generally thought to be safe. According to an epidemiological study, increased use of probiotic Lactobacilli products in Finland did not increase the incidence or frequency of bacteremia. Sporadic cases of bacteremia have been linked to specific lactobacilli strains in certain circumstances. Bifidobacteria are thought to be beneficial for breastfed babies' growth and development due to their prevalence in their digestive systems (Gwee et al., 2018). Because there have been few reports of Bifidobacteria causing host harm, their safety has not been called into question.

The recent use of large quantities of intestinal bacteria isolates in critically ill patients has sparked controversy. Only probiotic bacteria strains and applications that have been proven to work in treating illness should be used. A multidisciplinary approach is required to evaluate probiotic strains' toxicological, immunological, gastroenterological, pathological, and infectious effects. Because probiotics are designed to thrive in the human colon and provide health benefits, conventional toxicology and safety testing are insufficient. Several methods for determining the safety of LAB have been developed through in vitro, animal, and human clinical studies. The inherent properties of probiotic bacteria and their interactions with their hosts, can be used to assess their safety. Each potential strain's acute, subacute, and chronic toxicity must be assessed when ingesting extremely high doses of probiotic bacteria. Such testing may not be required for strains with a documented history of use.

As a result, the following should be included in probiotic bacteria safety considerations:

- Antibiotic resistance strains
- The ability of animals with weakened immune systems to spread disease
- Probiotic bacteria should be studied for toxin production. The EU Animal Nutrition Scientific Committee has proposed a possible strategy for assessing toxin production.
- Acting as a hemolysin
- Reactions occur within living organisms (D-lactate, bile salt de-conjugation).
- Physiological and genetic factors

In epidemiological studies, the incidence of adverse consumer events is monitored (post-market).

There are several labeling requirements.

For consumers to make informed decisions, accurate labeling and health claims are required. Each country's food law requires labeling food product, but the label must include all relevant information. Many countries allow only general probiotic health claims, but specific health claims should be permitted if sufficient scientific evidence exists (Gogineni et al., 2013). A probiotic would be more accurate to say that it "reduces the incidence and severity of rotavirus diarrhea in infants" rather than "improves gut health."

Such claims should be permitted on the label and in advertising. The manufacturer is responsible for accurately and truthfully labeling health claims.

Figure 8.2: Non-GMO labels.

As a result, the label must include the following information:

- The microbial species must be described to identify a probiotic present in food. When naming genera, species, and strains, nomenclature should be followed. Because all probiotic effects are strain-specific, the strain's name must also be mentioned if a selection process was used. Names should not be used to mislead customers about the potential health benefits of a strain.
- Each probiotic strain must contain at least the minimum viable amount at the end of the product's shelf life: To be included on the label, probiotic bacteria must be precisely measured in food products. The shelf life of each probiotic supplement and the number of viable bacteria remaining at the time of expiration, should be noted on the label (Gupta, 2011). The minimum efficacy level of each probiotic strain must be supported by scientific research and maintained throughout the product's shelf life.
- The number of probiotic bacteria in each serving is required to support the health claim.
- To the extent permitted by law, scientific data supports a detailed account of the physiological effect.
- Proper storage conditions, such as the product's recommended temperature.
- The phone number and email address of the company for customer inquiries.
- The recommended use conditions are risk-free.

- Food label information should not lead consumers to believe that an ingredient, nutritional component, or food can treat, relieve, cure, or prevent disease.

8.2 BIFIDOBACTERIUM ANIMALIS

B. animalis is an anaerobic, Gram-positive, rod-shaped bacterium found in humans and most other mammals. It is one of many bifidobacteria species that can be found in the digestive tract. Previously, *B. animalis* and *B. lactis* were thought to be separate species. The Bifidobacterium family includes two subspecies: *B. lactobacillus* and *B. lactis*.

The names *B. animalis* and *B. lactis* continue to appear on product labels due to their widespread use as probiotics. It is not always clear which subspecies of a species were used to create a product.

The most researched Bifidobacterium probiotic strain is BB-12® (BB-12®). The condition is described in over 300 scientific papers, including over 130 papers on human clinical studies. The genome sequence of BB-12entire ® has been deciphered and made public. BB-12® is derived from Chr. Hansen's dairy culture collection is highly stable in food and freeze-dried powders. The strain's characteristics and mechanisms have been determined through extensive in vitro testing. BB-12® has important probiotic properties such as bile salt hydrolase and strong mucus adhesion. Pathogen inhibition, improved barrier function, and immunological interactions are all BB-12 mechanisms ®. BB-12® has been shown to be beneficial to the immune system and gastrointestinal health (Gao *et al*., 2022). Clinical studies have shown that this product can survive the digestive process and help to maintain a healthy microbiome in the gut. In addition to these advantages, BB-12® has been shown to improve bowel function and protect against diarrhea.

Clinical studies have shown that BB-12® can reduce the frequency of acute respiratory tract infections while increasing the body's resistance to common respiratory infections. *Bifidobacterium animalis* subsp. lactis BB-12 reduces obesity in two ways when administered to rats exposed to human microbiota.

People generally believe probiotics are safe for human consumption, so taking them is a common way to influence the gut microbiota composition. Probiotics have been shown in most studies to be an effective treatment for maintaining intestinal microbiome homeostasis. Bifidobacterium probiotics are widely used because they can prevent and treat many gastrointestinal disorders in humans and animals, such as intestinal infections and irregularities in colonic transit. The ability of Subsp. animalis (BB-12) of *B. lactis* (BB-12) to tolerate gastric acid and bile, as well as adhere to the mucus membrane, is an essential probiotic feature. BB-12 has been given to babies, children, adults, and the elderly in successful clinical trials. Some reported benefits include treating infantile colic, boosting the immune system, lowering the risk of infant infection, and restoring the disturbed gut microbiota balance caused by severe acute malnutrition.

Around the world, BB-12 is also found in infant formula, fermented milk products, and nutritional supplements. Few studies have looked into the role of BB-12 in weight loss. The effects of several probiotic strains on obesity have been confirmed and documented, including weight loss, BMI reduction, and fat percentage reduction with probiotic supplementation (Faujdar et al., 2016). Probiotics derived from Lactobacillus and Bifidobacterium (*B. infantis, B. longum*, and *B. breve* B3) have been used successfully to treat animal obesity and its associated medical complications. When taken as a supplement, *Lactobacillus rhamnosus* GG has been shown to reduce weight gain in children. These studies have shown that probiotics are a safe and effective way to treat obesity.

8.3 BFIDOBACTERIUM BREVE

B. breve is a bacterium that belongs to the Bifidobacteria family. These organisms have a rod shape, are Gram-positive, and are anaerobic.

B. breve is frequently found in babies' gastrointestinal tracts, particularly those who have been breastfed. Breast milk has recently been found to contain B. breve strains. This may account for their abundance in these sources. Prebiotic HMOs are highly selective and can be found only in human milk. Unlike traditional prebiotics, these substances can only be metabolized by a few Bifidobacterium species.

During HMO fermentation, *B. breve* strains produce short-chain fatty acids, which aid in maintaining a healthy gut environment. Lactose, as well as the starches and sugars found in milk, are digested by *B. breve* enzymes. Natural digestion and nutrient absorption improve as a result. As a result, *B. breve* strains for babies and small children are popular. Several strains of B. breve have been tested on children and found to be effective for a wide range of medical conditions. Bifidobacterium breve M-16V® is a strain. *B. breve* and other "good" bacteria not only aid digestion but also protect the body from harmful "bad" organisms and improve nutrient absorption (Ezema, 2013). Despite a lack of scientific evidence, people use *B. breve* to treat various conditions, including blood infections, age-related memory, cognitive decline, and many others. It is highly improbable to work. Alzheimer's disease and other aging-related memory loss oral administration of *B. breve* does not appear to improve memory or cognition in older adults.

B. breve's probiotic benefits are best known for infants and children under two. Infant populations are much more prevalent than adult populations. Breve was discovered for the first time in the excrement of breastfed infants in 1990. Infants must consume *B. breve* to digest breast milk. In breast milk, human milk oligosaccharides are indigestible (HMO). *B. breve, Lactobacillus acidophilus*, and *B. bifidum* all help with pediatric constipation. *B. breve* is found in kombucha, water kefir, and raw sauerkraut.

8.3.1 The Skin Is Getting Better

Figure 8.3: Healthy skin illustration.

The ability of *B. breve* to improve skin health is one of its most valuable properties for adults. After drinking fermented milk containing *B. breve,* young women's skin and digestive health improved. Probiotic users had clearer skin than those who received a placebo. Drinking probiotic milk regularly, according to the researchers, can improve gut health and toxin elimination and help prevent dry skin (Cammarota *et al.*, 2014). It has also been shown in mouse studies to increase skin flexibility, prevent water loss, and reduce the damage caused by prolonged exposure to UV rays.

8.3.2 The Swelling Has Subsided

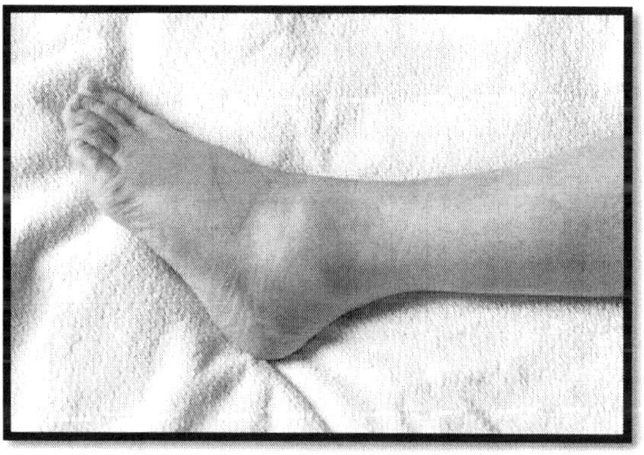

Figure 8.4: Swelled leg.

B. breve's anti-allergy properties may be especially beneficial for children. Anti-inflammatory responses are induced in premature infants, which may help prevent future allergic reactions. In infants, *B. breve* has been shown to reduce atopic skin reactions and cow's milk allergies. Another study found that children who took *B. breve* were less likely to develop eczema and atopic sensitization.

Furthermore, *B. breve* strengthens the immune system, reducing the symptoms of allergic rhinitis. The use of *B. breve* regularly reduces the body's immune response, resulting in fewer allergic reactions. Asthmatic patients benefit most from this (Deng *et al.*, 2020). Airway inflammation causes allergy attacks in asthma patients, which *B. breve* can help to alleviate. When combined with specific fructooligosaccharides, it may be more effective.

8.3.3 Gastroins are controlled by Both Children and Teens

B. breve has a significant advantage in the fight against obesity. According to one study, taking supplements can significantly reduce fat mass in adults prone to obesity. It may also help lower serum insulin, fasting glucose, and total cholesterol levels."

B. breve, like many other probiotics, helps children avoid diarrhea and constipation. It promotes frequent bowel movements and relieves stomach pains in children with functional constipation. The risk of dehydration is reduced when *Streptococcus thermophilus* is used to treat acute diarrhea in young children.

With the help of *B. breve*, young children with low birth weights can acquire healthy Bifidobacteria and normal intestinal flora. This also aids in the development of healthy body weight in infants.

B. breve is a probiotic bacterium that is extremely beneficial to the human body (*B. breve*). It is one of the few probiotic bacteria capable of outcompeting harmful bacteria due to its ability to digest many molecules. Breast-fed babies have a stronger immune system than formula-fed babies for various reasons, including this one. The primary digestive tract functions of *B. breve* are sugar fermentation, lactic acid production, and acetic acid production (Deshpande *et al.*, 2011). Sugar fermentation, lactic acid production, and acetic acid production are the digestive system's primary functions of *B. breve*.

8.3.4 Bifidobacterium breve Can Help Avoid Digestive Problems

Allergies, gas, irritable bowel syndrome, and diarrhea caused by B. breve deficiency have all been linked to antibiotics (IBS). *B. breve* appears to be a growth inhibitor for E. coli. Compared to other probiotic bacteria, it has been found to aid digestion because it can break down plant fibers, which are typically considered indigestible.

According to one study, *B. breve* promotes more frequent bowel movements in constipated children. According to the same study, the children reported less abdominal pain due to this.

B. breve concentrations in individuals can vary greatly. Some newborns have high concentrations of *B. breve*, while others have deficient concentrations, depending on whether the mother gives birth vaginally or via cesarean section. Even those born with a high *B. breve* population tend to lose some of it as they age. It will eventually account for less than 10% of the digestive-related bacteria in many adults, allowing less beneficial bacteria to overpopulate the gut during stressful times, resulting in illness or slowed recovery (Culligan et al., 2009). To keep the gut healthy, probiotic supplements should be taken regularly.

8.3.5 Bifidobacterium breve B-3's Effects on Fat Loss in Obese Adults

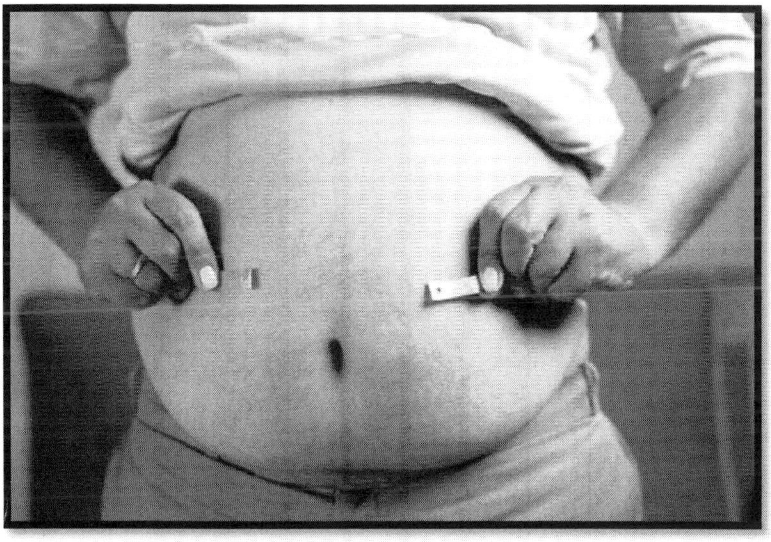

Figure 8.5: An obese person.

Obesity, hypertension, dyslipidemia, and diabetes are all caused by an abnormal or excessive fat accumulation. Although energy-dense food consumption and a low energy requirement for physical activity remain the primary causes of obesity, recent research indicates that the gut microbiota directly influences energy homeostasis and fat storage in adipocytes and strongly correlates with obesity. The metabolic endotoxemia hypothesis, which connects gut microbiota to chronic inflammation and other metabolic disorders, could explain the connection between gut microbiota and obesity. Obesity and metabolic disorders have been linked to gut barrier dysfunction in experimental animals and humans. Combined with a Western diet, this can potentially increase endotoxin translocation (Chalas *et al*., 2016). LPS, an outer membrane molecule found in Gram-negative bacteria, has been linked to inflammation, insulin resistance, and type 2 diabetes development. Sun et al. discovered that LPS-binding protein (LBP), a marker of subclinical endotoxemia, was higher in overweight/obese people than in normal-weight people (geometric mean 27.6 g/mL vs. 10.0 g/mL; p0.001). Bacteria supplementation improved gut integrity, reversed metabolic endotoxemia, and mitigated the effects of endotoxemia. Bifidobacteria, which may aid in the prevention and treating metabolic endotoxemia, can aid in the treatment of metabolic diseases.

Obesity has been linked to a higher Firmicutes-to-Bacteroidetes ratio in the microbiota. Contrary to popular belief, not all human trials have revealed the same gut microbiota composition in obese children. The prevalence of Bifidobacterium in the feces of obese subjects was significantly lower than that of lean subjects. A meta-analysis found that the gut microbiota of obese people consistently and significantly lacks bifidobacteria. Probiotic bacteria strain Bifidobacterium is the most effective at reducing obesity (Di Cerbo & Palmieri, 2015). Although Bifidobacterium spp. has been shown to have species- or even strain-specific anti-obesogenic effects, the strength of these effects is known to be influenced by the human context. Clinical studies with a specific strain on clearly defined subjects are required to determine the anti-obesity effects of probiotics.

Bifidobacterium breve B-3 is undoubtedly, one of the most effective anti-obesogenic strains available. We discovered in a previous study that giving this strain to diet-induced obese mice reduced weight gain and visceral fat deposition while increasing total serum cholesterol, glucose, and insulin levels in a dose-dependent manner. A daily dose of 50 billion CFU of lyophilized *B. breve* B-3 powder reduced human body fat mass; however, some subjects were taking diabetes, high blood pressure, or high cholesterol medications.

8.4 BIFIDOBACTERIUM LACTIS

Bifidobacterium lactic intestines contain these probiotic bacteria (B. lactis). Lactic acid and acetic acid are produced during the fermentation process.

By incorporating beneficial bacteria, such as *B lactis*, into the diet, digestion can be improved, nutrient absorption, and defense against disease-spreading "bad" bacteria. *B. lactis* is a subspecies of *B. animalis*. This substance is naturally present in the human digestive system and is used in probiotic supplements.

B. lactis treats respiratory infections, constipation, IBS, and excessive crying in infants (colic). This plant can be used to treat hay fever, dysentery, and a variety of other ailments. *B. lactis* cannot be used to treat COVID-19 because there is no evidence to back this up (Daliri & Lee, 2015).

Another name for *Bifidobacterium lactis* that may be seen on product labels is *Bifidobacterium animalis subsp. lactis* (or *B. lactis*). Other fermented foods and probiotics, such as yogurt, kefir, and fermented milk, do not contain *B. lactis*. They cannot be used interchangeably.

Sounds produced by a crying baby (colic). According to a new study, taking *B. lactis* orally alleviates colicky babies' symptoms .

Constipation that persists. Constipation can be relieved by taking *B. lactis.*

Constipation and lower abdominal pain are symptoms of a long-term intestinal disorder (irritable bowel syndrome or IBS). *B. lactis*, when taken orally, can help relieve IBS symptoms. Supplementing probiotics with other probiotics has not been proven beneficial (Dai et al., 2018).

Respiratory tract infection. Taking *B. lactis* by mouth can help prevent respiratory infections like the common cold in healthy people. On the other hand, those with more serious medical problems appear unaffected.

Vacuum-cleaning agents Children who consume *B. lactis* appear to be unprotected against cavities.

B. lactis is used to keep blood sugar levels stable and to maintain a healthy weight.

A new study found that *B. lactis* reduced fat mass and glucose intolerance in obese and diabetic mice. One theory was that a decrease in the pro-inflammatory lipopolysaccharides (LPS) from bacteria associated with leaky gut was to blame. An imbalance in gut bacteria has been linked to various conditions and diseases, making this an excellent example of how a "good guy" can be introduced into the ecosystem and help shift things in the right direction. In this case, the probiotic aided weight loss and blood sugar control.

It has been demonstrated that *B. lactis* HN019 has significantly impacts people suffering from metabolic syndrome. If the study's findings are confirmed, probiotic supplementation should be given more consideration, according to the study's authors, who also noted that this probiotic had positive effects on measurements of nitric oxide metabolites, antioxidants, and inflammation. Including a specific strain of probiotics in the diet can significantly impact the health (Chugh & Kamal-Eldin, 2020). As a result, it demonstrates the wide range of health issues that can be caused by an imbalance of good and bad bacteria in the digestive tract.

8.4.1 Bifidobacterium lactis, Intestinal Health, And the Immune System

This beneficial bug may help with immunity and intestinal health. The gut-specific immune response was altered in *B. lactis* (BB-12)-infected pigs, reducing the robust immune response that parasites may elicit. There has been much research and debate about the importance of the gut to the immune system.

The gut microbiome, in our medical opinion, has a significant influence on the immune system. The studies I have discussed, as well as numerous other published studies, show a strong correlation between gut microbiome composition, inflammation, and immune reactivity. A healthy microbiome is important for various reasons, including this one. A healthy digestive system boosts immunity! According to a recent study, *B. lactis* and other strains of bacteria in a probiotic formulation significantly reduced inflammation in the body. According to this new study, changes to gut microbiome can improve one's health and immune system.

8.4.2 B. lactis and Heart Health

Probiotics can boost the immune system and reduce inflammation after a heart attack. According to a study published earlier this year, supplementing with *B. lactis* 420 (B420) reduced the effects of a heart attack. As we have long suspected, gut and heart health are inextricably linked. If the gut microbiome is well-balanced and contains enough *B. lactis* and other beneficial bacteria, distant tissues and organs may benefit (Choudhary & Tandon, 2009).

One of the most extensively researched probiotics on the market is *B. lactis* BB-12. This probiotic's numerous health benefits, including its ability to fight pathogens, strengthen the gut lining, and boost the immune system, have been discussed in dozens of publications. Antibiotic use can reduce side effects while also promoting a healthy gut microbiome.

Foods that may contain the bacteria *Bifidobacterium lactis*.

- Yogurt with Greek origins
- Cheddar cheese (optional)

A variety of soft cheese

A cup of sour cream

Which foods contain Lactobacillus lactis?

Bifidobacterium lactis is a starter culture in yogurt, cheese, cottage cheese, buttermilk. Lactation fluid contains *Bifidobacterium lactis* bacteria. Infant formulas may contain *Bifidobacterium lactis spp.* ingredients. This probiotic can be successfully added to a wide range of foods, including infant formula, because it can withstand freezing and drying.

How Does *Bifidobacterium lactis* Work?

As Mary Poppins says, consistent hard work can sometimes lead to magical places. *Bifidobacterium lactis* is a prime example. *Bifidobacterium lactis* aids vitamin and mineral absorption in the intestines and digestion by producing folic acid (vitamin B9), transforming minerals into absorbable forms, and being essential for the cellular gateways that allow minerals into cells. Despite its friendliness, *Bifidobacterium lactis* must not be overlooked! Gliadin, found in Celiac disease, has been shown to have anti-toxic effects (Zhang *et al.*, 2016). These probiotic species mitigate these effects. It also protects the lining of the digestive tract from pathogens.

8.5 BIFIDOBACTERIUM LONGUM

In combination, an oral probiotic supplement containing the strains *Bifidobacterium longum*, *Lactobacillus acidophilus*, and *Enterococcus faecalis* may provide anti-diarrheal and immunomodulatory properties. Beneficial bacteria in *Bifidobacterium longum/Lactobacillus acidophilus/ Enterococcus faecalis* probiotic supplements may alter the normal GI microflora after oral ingestion, improving digestion and aiding in colonization maintenance. They produce lactic acid, hydrogen peroxide, and other compounds that inhibit pathogen growth during fermentation. The GI tract's mucosal epithelium may be protected from toxins and other chemicals by a protective intestinal barrier formed by bacteria during colonization. This helps to keep bacteria and infections at bay. This substance can potentially inhibit inflammatory cytokines while also improving innate and acquired immunity. Combining these effects can reduce diarrhoea, protect the gastrointestinal mucosa from damage, improve digestion, and reduce inflammation.

Bifidobacterium longum is required to keep the body healthy. *Bifidobacterium longum* bacteria in the intestine and genital tract aid digestion and the immune system. It is safe to consume these Gram-positive anaerobic bacteria. They are grown at 37 °C in a medium containing Reinforced Clostridia. The genome of *Bifidobacterium longum* contains approximately 60% Guanidine-Cysteine. This microorganism's genome contains 2.26 gigabytes of data. *Bifidobacterium longum*, a bacterium known for fermenting sugars into lactic acid, is found in yogurt and other fermented foods. As a result of this action, the pH of the small intestine rises. This behavior demonstrates *Bifidobacterium longum's* probiotic properties (Yao et al., 2020). Probiotics are microorganisms that benefit their hosts. Scientific research is currently being conducted to isolate genes from B. longum. An important part of this research is determining how this bacteria can survive in the digestive tract and other harsh environments.

Bifidobacterium longum is essential for human development in the early stages. Breastfeeding can transmit *Bifidobacterium longum* from mother to child. This early transfer strengthens the infant's immune system. Allergies and diarrhea are less likely in breastfed babies. *Bifidobacterium longum* protects adults from cancer, including colon cancer, by balancing the microbes in the gut. Vaginitis, or vaginal infection, can be avoided with the help of this bacterium. Lactic acid, which *Bifidobacterium longum* produces, reduces the risk of vaginal vaginitis.

Because of its gene pool, *Bifidobacterium longum* can synthesize approximately 19 amino acids. People can be used to test applications. Researchers are investigating the relationship between constipation, weight loss, and obesity. According to the evidence, people who took *Bifidobacterium longum* had better weight loss and constipation results than those who took laxatives.

Bifidobacterium longum is being studied as a gene delivery vehicle for cancer treatment in Japan. In rat studies, *Bifidobacterium longum* was found to be able to penetrate the tumor barrier and accumulate within the tumor. This new study could help scientists understand how to deliver anti-cancer drugs directly to the source of the disease (Wescombe *et al.,* 2012).

Scientists discovered this important microbe in feces over a century ago. *Bifidobacterium longum* can convert many sugars into lactation, which aids digestion and disease prevention. Fortunately, *Bifidobacterium longum* has been extensively researched to understand its role in the human body better, and this research will continue so that we can all benefit from the findings.

Because of their "Y" shape, *Bifidobacterium longum* is thought to be rod-shaped bacilli. Although their distinctive "Y" shape increases the surface area to volume ratio and may aid in the compartmentalization of cellular structures, its function is unknown. *Bifidobacterium longum* is naturally found in the human vaginal and digestive tracts. Because *Bifidobacterium longum* are anaerobic microorganisms, hypoxic conditions are likely to exist in both environments. *Bifidobacterium longum* is anaerobic because it lacks the catalase enzyme, found in aerobic organisms. 'Bacteria with peptidoglycan cell walls, such as *Bifidobacterium longum*, have a single cell membrane. *Bifidobacterium longum* can reach a length of 8 m.

Infants have higher levels of *Bifidobacterium longum* in their intestines than adults. While *B. longum* accounts for only 3% of the microbiota in adults, it accounts for 9 out of 10 bacteria in a baby's stool sample. The unique oligosaccharide metabolism of *Bifidobacterium longum* can explain this. A large number of genes encode the oligosaccharide-catabolizing proteins produced by *Bifidobacterium longum*. An infant's nutritional needs are entirely met by his or her mother's milk. A wide variety of oligosaccharides unique to breast milk are abundant. It makes sense to supplement infants' diets with *Bifidobacterium longum* to promote normal growth and development. Oligosaccharides are short carbohydrate polymers composed of three to ten simple sugars linked together by oxygen or nitrogen bonds (Valero-Cases *et al.*, 2020). Oligosaccharides are found in plant fibres and breast milk. Because *Bifidobacterium longum* is present in the vagina, infants can contract it through the birth canal. Our digestive systems require a broader range of microorganisms as we age and consume a wider variety of foods. In vitro studies, on the other hand, show that *Bifidobacterium longum* can prevent IECs from developing a leaky gut. In vivo studies have revealed that *Bifidobacterium longum* has a beneficial effect on the digestive system.

8.6 LACTOBACILLUS REUTERI

This species inhabits the mouth and the gastrointestinal tract. According to one study, the bacteria that cause tooth decay are reduced. It may also be beneficial to the digestive system.

Lactobacillus reuteri is a rod-shaped, anaerobic Gram-positive microbe. These heterofermentative lactic acid bacteria are found in the intestines of mice, pigs, chickens, humans, and many other animals and plants. Furthermore, they can be found in human breast milk. *Lactobacillus reuteri* lab cultures thrive on MRS media at 37 degrees Fahrenheit. They have also been discovered to thrive in biofilms. According to the United Nations Food and Agriculture Organization, probiotics are "living microorganisms that confer a health benefit on the host when administered in adequate quantities." This concept was first proposed by Elie Metchnikoff in the early 1900s. Produces reuterin antimicrobial, inhibiting pathogenic bacteria, fungi, and protozoa growth. According to current research, *Lactobacillus reuteri* can treat infantile colic, eczema, and *Helicobacter pylori* infection. This is due to the presence of beneficial bacteria.

When *Lactobacillus reuteri*, a Gram-positive rod, is cultured, it forms chains but does not produce endospores. Some strains can also produce biofilms as thin as 5-7 m. The host-associated, mutualistic microorganism *Lactobacillus reuteri* thrives in animal digestive tracts (Tadesse, 2012). A facultative mesophilic anaerobic anaerobe prefers acidic environments and requires the presence of a host (for). Glucose fermentation in *Lactobacillus reuteri* produces carbon dioxide, ethanol, acetate, and lactic acid. It can also anaerobically degrade glycerol, producing the antibacterial reuterin (3-hydroxypropionaldehyde). Researchers discovered that the bacterium *Lactobacillus reuteri* produces vitamins B12 and folate, which are required by animals and humans.

Lactobacillus reuteri strains have been found in animal gastrointestinal tracts. Because *Lactobacillus reuteri* has only been studied as a probiotic, it has been impossible to determine whether it is harmful. *Lactobacillus reuteri* ferments glycerol to produce reuterin. Reuterin, an antibiotic with a broad spectrum that kills Gram-positive and Gram-negative bacteria, can kill most intestinal pathogens. According to research, reuterin from *Lactobacillus reuteri* is effective against pathogenic strains of *E. coli*, Clostridium, *Salmonella enterica*, and *Vibrio cholera*. Even though reuterin production is well understood, the mechanism by which it inhibits the growth of such a diverse range of pathogenic microorganisms remains unknown. *Lactobacillus reuteri* inhibits other bacteria's colonization by colonizing epithelial cells, contributing to the antibacterial properties of *Lactobacillus reuteri*. *H. pylori*, the most common cause of gastric ulcers, may be avoided by competing *Lactobacillus reuteri* strains for colonization.

Lactobacillus reuteri has been tested for the production of folate and vitamin B12 as a possible supplement. Low levels of vitamin B12 in the diet have been linked to cardiovascular disease, anemia, and cancer. Vegetarians are more likely to be deficient in vitamin B12, so supplementing with *Lactobacillus reuteri* would be beneficial. *Lactobacillus reuteri* was initially used to treat necrotizing colitis, a digestive disorder characterized by infection and inflammation that can be fatal for newborns, particularly premature babies. Because of its anti-inflammatory properties, *Lactobacillus reuteri* was chosen. ATCC 55730 and DSM 17938 bacteria were used in research on necrotizing colitis (Smolyansky, 2010).

The popularity of *Lactobacillus reuteri* grew following the discovery that changes to the digestive system can influence the immune system. In animal studies, the *Lactobacillus reuteri* ATCC PTA 6475 strain was found to increase oxytocin and testosterone levels and skin quality. Furthermore, animal studies have revealed potential benefits for bone density, hair quality, and weight loss prevention from obesity-promoting diets.

Humans have used lactic acid bacteria for millennia for millennia to preserve and improve the nutritional value of delicate foods. *Lactobacillus reuteri* boosts the number of intestinal Treg cells. Many foods now contain Bifidobacteria and lactobacillus strains.

Numerous strains of *Lactobacillus reuteri* have been shown in studies to be beneficial for oral health and various gastrointestinal disorders. Furthermore, Lactobacilli are thought to contribute to the development of allergies.

8.6.1 The Colonization of the Digestive Tract by Lactobacillus reuteri

Lactobacillus reuteri colonies can be found in both human and animal guts. Its probiotic properties are thought to be due the colonization of the gastrointestinal (GI) tract. To provide the best conditions for digestion and absorption, certain areas of the digestive system have evolved hostile environments for microorganism colonization and proliferation (Abuajah *et al*., 2015). A microorganism must first survive these sometimes-harsh conditions before colonizing the gut. Many strains of *Lactobacillus reuteri* are resistant to bile salts and low pH, allowing them to colonize and survive in the digestive tract.

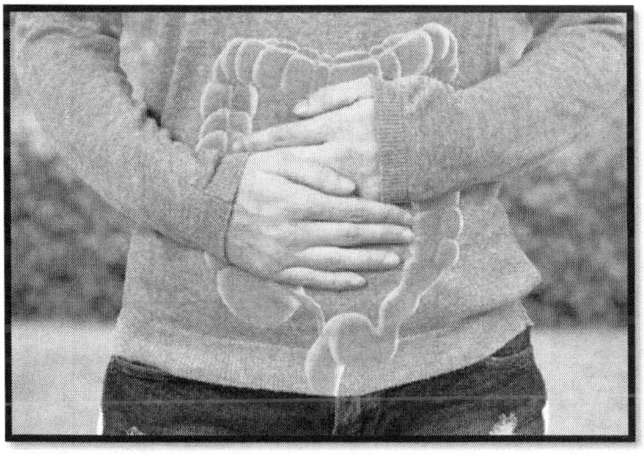

Figure 8.6: A digestive system.

8.6.2 Metabolite Production Profile of Lactobacillus reuteri

The metabolite profile of *Lactobacillus reuteri* is associated with antimicrobial and immune-boosting properties. A metabolite is a substance that the body produces or uses to digest food, chemicals, and medications. The body generates the energy needed for growth, reproduction, and health maintenance through a process known as metabolism. These four *Lactobacillus reuteri* metabolites are good for the health.

Histamine functions as a neurotransmitter-

In some strains, *Lactobacillus reuteri* can convert L-histidine to biogenic amine histamine. The biogenic amines are derived from the amino acid building blocks of proteins. The most well-known biogenic amine is histamine. It is produced by the body and plays an important role in allergic reactions.

According to one study, histamine derived from the *Lactobacillus reuteri* strain 6475 suppressed tumor necrosis factor (TNF). TNF is a pro-inflammatory protein in the body (Aggarwal *et al*., 2013). This suggests that specific bacterial strains may effectively reduce or eliminate inflammation in autoimmune diseases.

Minerals and Vitamins-

You require 13 essential vitamins because your body cannot produce the remaining vitamins. Various strains of *Lactobacillus reuteri* can produce a variety of vitamins, including B12 (cobalamin) and B9 (folate) (folate). B12 is required for proper nervous system function and the production of DNA and red blood cells. The vitamin also affects the production of reuterin.

Exopolysaccharide (EPS) is a type of polysaccharide (EPS)

Exopolysaccharides (EPS) are long-chain polymers that were developed and produced primarily by bacteria and microalgae. EPS is required for the formation of biofilms and the attachment of *Lactobacillus reuteri* to body surfaces. *Lactobacillus reuteri* generates EPS (Stavropoulou & Bezirtzoglou, 2020).

A biofilm is a microbial community that can colonize a variety of surfaces. Microorganisms that can form biofilms include fungi and bacteria. The EPS from *Lactobacillus reuteri* can also prevent *E. coli* from adhering to epithelial cells.

8.6.3 The Role of Lactobacillus reuteri in the Treatment of Leaky Gut

If you've ever had stomach cramps or bloating, you know how inconvenient it can be. Digestive issues, in addition to acne or migraines, may indicate a leaky gut. Increased intestinal permeability, also known as "leaky gut," is a digestive disorder in which bacteria and toxins enter the body through the intestinal wall. The functionality of the gut barrier can be influenced by physical, biochemical, and immunological barriers. Increased intestinal permeability can result in a leaky gut due to abnormalities in the intestinal barrier (Aponte *et al.*, 2020). Numerous studies have shown probiotics strengthen the intestinal wall by improving mucosal barrier function. *Lactobacillus reuteri* is one of the most extensively researched probiotics. This minuscule bacterium helps to heal the leaky gut by sealing the holes in the intestinal lining that caused it.

8.6.4 The Effect of Lactobacillus reuteri's Candida

Candida yeast can be found in trace amounts on the skin, mouth and, intestines, among other places. It is not a problem at normal levels. Candida, on the other hand, can cause candidiasis if it begins to proliferate uncontrollably. Candida is the most common fungus that causes infections in humans. According to research, *Lactobacillus reuteri* may have antifungal properties, allowing it to eradicate various Candida species by preventing their growth,.

Two strains of *Lactobacillus reuteri*, DSM 17938 and ATCC PTA 5289, completely inhibited the growth of Candida species, including Candida albicans, in investigating the bacteria's antifungal potential. *Candida albicans* is a common Candida species in the intestinal flora. *C. albicans* infections include oral thrush, genital yeast, and urinary yeast infections.

Lactobacillus reuteri, a probiotic organism, is naturally present in human breast milk. Lactobacilli, according to research, maybe a reliable indicator of a healthy, antibiotic-free milk microbiome. Breast milk may provide colonizing bacteria to a newborn because it contains various nutrients and a unique collection of bacteria (Saraf *et al.*, 2010).

Colic condition is characterized by inconsolable crying that begins during the second week of life and lasts for about three months. Colic affects up to 28% of infants, causing understandable concern among parents and medical professionals. Crying is the most common cause of distress in the first three months of a baby's life. Breastfeeding mothers have received similar advice over time. Although studies have shown that Lactobacillus treatment is superior to simethicone treatment, they suggest that low Lactobacillus numbers in the intestine may contribute to colic. Simethicone is an anti-gas medication that relieves bloating and helps gas pass through the body.

CHAPTER 9

HOW PROBIOTICS FUNCTION IN THE HUMAN BODY

CONTENTS

Probiotics—live microorganisms given to the host to improve their health—are becoming more popular in the food and nutrition industry. Probiotics influence the populations and activities of host microflora in the skin, vaginal tract, and gastrointestinal tract. Before this discovery, the importance of gastrointestinal flora to human health had been grossly underestimated. Although the concept of colonization resistance has been around for a long time, the medical community has only recently recognized its significance.

The normal flora inhibits bowel colonization by potentially pathogenic exogenous and endogenous microorganisms. "The human suffering that results from infection," it says instructional, "especially in the case of infections that can be easily eliminated through antimicrobial agents, has spawned clinical attitudes that view these microorganisms as hostile disease agents that must be completely eradicated." In the relationship between multicellular organisms and microbiomes, synergy, not hostility, should be considered (Sistek *et al.*, 2006).

As a result, preserving rather than removing the natural flora may be better for the patient's health. Probiotics are more important than simply acknowledging the benefits of the host's native flora because exogenous microbes can use them to benefit the host. This concept was popularized by Russian immunologist Elie Metchnikoff, who received the Nobel Prize in the early twentieth century for his work. Probiotics are now available as dietary supplements, pharmaceuticals, and food for human and animal consumption.

Why should we consider probiotics because of their significance to human health? The gastrointestinal tract and the microorganisms that live there have a wide range of effects on human health. Pathogens can only infect a limited number of organisms in most micro-fora. By increasing the body's resistance to infection, antibiotic use (and the spread of antibiotic-resistant pathogens) and the severity of rare but severe secondary infections can be reduced. Inflammatory bowel diseases such as Crohn's disease and ulcerative colitis are incurable and may necessitate colon removal surgery due to a breakdown in intestinal-fora tolerance. sSome intestinal bacteria's abnormal behavior may produce carcinogens, which aid in the spread of colon cancer (Sidhu *et al.,* 2015). By exposing the intestinal immune cells to the appropriate microbes during infancy, allergies can be avoided later in life. These findings imply that microbial interventions could be beneficial to health.

The World Health Organization defines probiotics as live microorganisms beneficial to the host's health when administered in adequate amounts. To encompass its use in a wide range of body sites and medical conditions, the term "probiotics" is intentionally broad in scope. However, the term "probiotic" should never be used to describe a living organism unless its health benefits are proven. The genomes of these microorganisms must be classified and sequenced based on strain.

9.1 DO PROBIOTICS HAVE ANY ADVANTAGES?

When probiotics are consumed, which maintain the balance of the intestinal flora, pathogenic microorganisms are less burdened. The immune system can be calmed and inflammation can be reduced by restoring a healthy balance of bacteria in the digestive tract. Improved digestive health benefits overall health by improving energy, mood, cognitive function, and hormonal balance. Numerous studies have shown probiotics to improve and maintain human health (Singh *et al.*, 2011). Probiotics have been demonstrated to be effective in the treatment of gastrointestinal disorders such as infectious and antibiotic-induced diarrhea. Probiotics may aid in weight loss, insulin resistance, depression and anxiety, and even fewer and shorter flu bouts.

Given the wide range of favorable health outcomes, it is reasonable to wonder how this is possible. How can we benefit from the consumption of probiotic microorganisms in terms of our health? The key to unlocking this mystery is our microbiome. There are trillions of microorganisms on our skin, in our mouths, and all over our bodies. These microorganisms, dubbed the human microbiome, have been shown to influence our nervous, immune, and metabolic systems.

Numerous microorganisms have been discovered in our bodies over the years, some of which have proven to be dangerous. Intestinal microorganisms can convert cellulose into short-chain fatty acids, which our colonic tissues can use for energy and also help prime the immune system toward a more balanced state of health. Microorganisms that produce endotoxin, an inflammatory substance, are more dangerous. This suggests that certain microorganisms can significantly affect our bodies when consumed or applied as probiotics and that some microorganisms are better suited to be beneficial than others (Singh & Natraj, 2021). Because probiotics have the potential to have a long-lasting effect at the site of action, a measurable response can be achieved even if the number of microorganisms present in probiotic foods or dietary supplements is significantly lower than the number of microorganisms in our microbiome.

Inquiring minds want to know about the health benefits of probiotics and what happens to the body when you take them. Probiotics can benefit health in a variety of ways, including:

- Exchange information with other microorganisms in the human microbiome.
- Encourage the growth of beneficial bacteria in the microbiome
- Prevent the growth of harmful bacteria in our microbiome
- Direct communication with internal organs such as the colon.
- Create medications to treat intestinal permeability syndrome or reduce inflammation.
- affect the immune system

Some probiotics work by altering our microbiome, among other things. Probiotics have the potential to influence the production and activity of microbiome bacteria. Even if the probiotic does not colonize for long, research shows that these changes can occur. Probiotic-induced changes in the human microbiome may influence how the microbiome influences organ function. Bifidobacterium and Lactobacillus probiotic strains, for example, produce organic acids and antimicrobial substances that inhibit the growth of potentially harmful bacteria that produce endotoxins in the intestine. Less of these harmful bacteria means less inflammation and degradation of barrier integrity (Rusu *et al.*, 2019). This mechanism is indirect because the resident microbiome determines the effectiveness of a probiotic at a specific body site.

On the other hand, immune, endocrine, and epithelial cells recognize probiotics and their secreted metabolites and other compounds. Once identified, a chain of events occurs, including the suppression of inflammatory responses and the repair of gut permeability. Probiotics' direct effects may cause long-term changes at the site where they are applied locally (such as the digestive tract) and in other parts of the body, similar to how probiotics alter the gut microbiome. It is important to note that not all conditions benefit from a specific probiotic. Even within the same species, microorganisms' genetic diversity allows for a wide range of physiological reactions. Several strains of the *Lactiplantibacillus plantarum* (previously known as *Lactobacillus plantarum*) can produce cytokines over a physiologically significant range comparable to that observed for other bacterial species and genera.

Probiotics' health benefits and applications in foods and beverages vary. Although predicting which strains will be the most effective is currently impossible, scientists are researching to identify the specific probiotic characteristics required for the observed health outcomes. Understanding the molecular mechanisms of probiotic function will ultimately aid in improving probiotic selection and usage recommendations, whether a probiotic affects mucosal tissues directly, indirectly by altering the human microbiome, or both. In the meantime, look at the labels of probiotic products to see which species and strains are present. It is also important to consider the microbes' ability to withstand different environmental conditions when using probiotics in foods or beverages. Refrigeration is usually required to keep Lactobacillus and Bifidobacterium strains alive. As a result, these microorganisms are appropriate for products that will be chilled during distribution and storage (Reid, 2016). These species are unlikely to tolerate high temperatures and acidic environments during processing. Endospore-forming strains can withstand a wider range of temperatures and pH levels due to their robust spore coats. Like a plant's seed, these strains remain dormant until the proper conditions (such as temperature, pH, and water activity) are met. When working with probiotics in foods and beverages, it is critical to consider the science, strain characteristics, and intended application.

The European Food Safety Authority recently rejected health claims made by commercially available probiotics due to a lack of evidence, citing one of the issues as poorly understood probiotic mechanisms of action. Furthermore, because some clinical trial authors failed to link microbiota analysis to the observation of an expected clinical effect, no investigation into the probiotic's potential impact on gut microbiota balance was conducted. Thus, the mechanisms of action of probiotics are primarily studied in vitro or using animal models, limiting the ability to extrapolate the findings to humans.

The first possible mode of action for probiotics could be to alter the host's microbiota. The "barrier" effect, also known as resistance to colonization, was one of the first proposed modes of action. It is used to prevent or limit the colonization of pathogenic bacteria. Bacterial inhibition could be caused by the production of broad-spectrum inhibitory bacteriocins, metabolites like short-chain fatty acids that cause a pH drop that makes bacterial growth less favorable, or antimicrobial biosurfactants. This barrier effect may employ two additional mechanisms in addition to competition for binding sites and inhibition of adhesion: competition for binding sites and inhibition of adhesion (Qi *et al.*, 2019).

The second mode of action focuses on improving the gut mucosa's barrier function. This barrier function is related to the effectiveness of tight junctions between intestinal epithelial cells. This barrier function is also supported by panteth cells, which produce antimicrobial peptides (defensins, lysozyme), and mucus cells, which act as a protective layer preventing direct contact with intestinal lumen bacteria. Probiotics can thus influence tight junction proteins by improving their physiological barrier function and signaling pathways that lead to defensin production or an increase in the mucus layer.

The third mode of action is immune system modulation. Gut-associated lymphoid tissue is found in the small intestine and contains more than 70% of immune cells. Antigens enter the body via Peyer patches, which have a follicular center and M cells attached to the epithelium. Immune response activation requires the recognition of specific innate immune cell receptors (epithelial cells, dendritic cells, and macrophages) (epithelial cells, dendritic cells, and macrophages). The majority of these pattern recognition receptors, are toll-like receptors (TLRs) and agents with nucleotide-binding oligomerization domains (NODs). Microbe-associated molecular patterns are structural elements repeated on the surface of microbes that interact with the gut epithelium and stimulate lamina propria immune cells. The result is T helper lymphocyte (Th) differentiation, activation of regulatory T cells, and the production of pro- or anti-inflammatory cytokines (Prins, 2011). Probiotic-potential bacteria, particularly lactic bacteria, may have various effects depending on the cytokine profile. The effects could be localized and limited to gut immunity stimulation (for example, stimulating secretory IgA production) or systemic.

Patients can benefit directly from probiotic strains that provide digestive-symptom-improving enzymes such as beta-galactosidase or other enzymes. Probiotic bacteria can exert their effects through several mechanisms, including:

- Their structural components include flagellin, peptidoglycan, LPS, and DNA (especially CpG motifs);
- They contain the following compounds in addition to their metabolites (particularly short-chain fatty acids): (especially short-chain fatty acids).

- They can affect the microbiota directly by colonizing the digestive tract or indirectly by increasing the inoculum of beneficial bacteria.

9.2 SIGNS YOU NEED PROBIOTICS

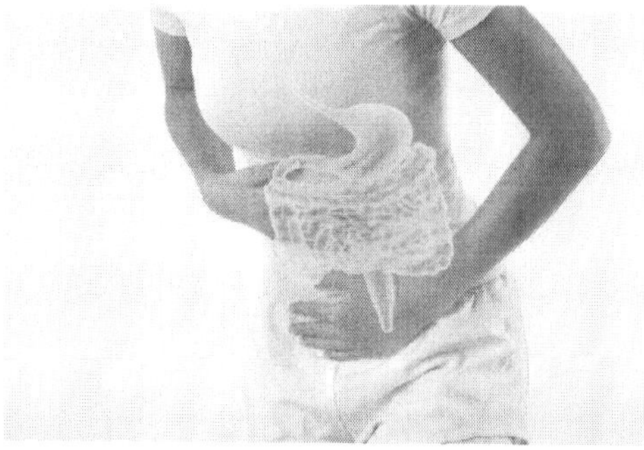

Figure 9.1: Gut discomfort illustration.

Gut health is critical to our overall health and happiness. Various vague health symptoms, such as fatigue, anxiety, joint pain, and brain fog, are thought to be related to gastrointestinal issues. One of the most effective ways to improve gut health is to consume probiotics. Probiotics are truly remarkable for their ability to treat many illnesses (Pratap *et al.,* 2020). They provide immune system support, hormone regulation, and digestive health, to name a few advantages. However, why are probiotics required?

Problems with health that do not improve with simple lifestyle changes are the best indicator that probiotic supplements may be required. Before dealing with long-term health issues, a solid foundation must be in place. This includes stress reduction, healthy eating, and regular exercise. Sleep is also an important factor to consider. If symptoms still persist after taking these crucial first steps, there may be a gut health imbalance, in which case probiotics can help.

If you have any of these symptoms, taking probiotic supplements may help;

- Skin problems, immune system boosts, mood disorders, and weight loss can all be attributed to gut health. Probiotic-rich foods and supplements can improve digestive function. This method may also help improve certain health conditions caused by gastrointestinal bacterial imbalances.

"Good" bacteria must be present in the gut to maintain a healthy digestive system, immune system, and ability to fight off harmful bacteria. Probiotics aid in vitamin and mineral absorption and can also help regulate mood-enhancing neurotransmitters in the brain. These are some reasons why taking care of one's gut health can help relieve stress, anxiety, and depression symptoms.

If probiotics are taken at the right time, they can help recover from chronic illness. However, if the necessary health-promoting factors are missing, probiotics may fail to provide the needed relief.

Although research indicates that probiotics can help with various health issues, this does not mean that the supplements are ineffective on their own. As a result, if there is a suspicion that a weakened gut causes the symptoms there is no "wrong" time to start taking probiotics (Pandey et al., 2015). A lack of healthy gut bacteria can cause asthma, skin conditions, infections, irritable bowel syndrome, and various other conditions. Could you benefit from taking probiotics? These symptoms may indicate a bacterial imbalance in the intestines.

- Although if antibiotics have not been taken in a long time, you did. Despite their ability to save lives, antibiotics have some drawbacks. By reducing the beneficial bacteria in the gut, they get rid of the bad bacteria that are causing problems in your body and kill them. Antibiotics can eliminate even "good" bacteria, regardless of whether they are considered "bad." You will likely need to rebalance your gut's good bacteria after taking antibiotics. Probiotics should always be kept on hand in case you need to retake antibiotics in the future. Research shows a link between age-related inflammation and altered gut flora, emphasizing the importance of gut repopulation.
- Asthma and allergies are major issues for you. Increase the number of good gut bacteria to treat the leaky gut syndrome, which can cause food allergies. Clinical studies show that taking probiotics and medication helps reduce asthma symptoms. Asthmatics and allergy sufferers should look into the effectiveness of probiotics (Arora & Baldi, 2015).
- At least one psychiatric condition is present. Because of the interaction between gut bacteria and brain neurotransmitters that control mood, bad gut bacteria can cause depression, anxiety,

irritability, and other mental health problems. If you have a history of depression or other mental health issues, you should consider taking probiotics and B vitamins.

- Your illness was caused by food poisoning. The bad bacteria may outnumber the good bacteria in the gut after a bout of food poisoning. Remember the last time you became ill from something and see if you have had any digestive issues. Constipation, diarrhea, and other digestive problems indicate that you should seek medical attention immediately.
- A low metabolic rate means not enough calories are being burnt. When Washington University School of Medicine researchers examined the gut bacteria of 154 people, they discovered an intriguing correlation. Obesity has been linked to a decrease in the number of bacterial species. It only takes two simple steps to replenish the gut with high-quality probiotics and nourish it with various plant-based fibers (Attri *et al*., 2021).
- You're constantly ill. Flu, colds, and urinary tract infections can all be aggravated by a lack of good bacteria in the intestines, weakening your immune system. Probiotics may be necessary to rebuild your immune system if you get sick frequently or run out of sick leave at work.
- You have psoriasis and acne on your skin. Acne, eczema, and psoriasis are frequently associated with digestive issues because waste elimination keeps the skin clear and healthy-looking. If topical treatments are not working, the skin may need to be treated internally with probiotics. The skin and digestive system may appear to be separate parts of the body, but they are, in many ways, intertwined. According to studies, a healthy digestive system can be seen on the skin. The microbiome that lives in our guts regulates the "gut-skin axis," which has been linked to digestive disorders. Another advantage of having a healthy microbiome is clear and glowing skin. The bacteria that live in our digestive tracts directly impact our health (Parvez *et al.*, 2006). Human strain probiotics and a fiber-rich diet are a winning combination for the digestive system. It will benefit the entire body!

- Your symptoms have no known cause. Probiotic supplements are frequently prescribed to treat functional gastrointestinal disorders and other difficult-to-diagnose symptoms.
- Type 2 diabetes and insulin resistance

Bacteria of various species have different dietary needs. Some people prefer the long-chain fibers found in vegetables, fruits, and other plant-based foods. Others favor refined flour and sugars like white sugar. Bacteria in the body can influence desires. Making wise food choices can significantly influence how much food the gut bacteria consume and grow. As a result, they have devised some novel approaches to achieving their objectives. Most of the time, this is done for the greater good. Because the body is their only home, you are the only food source for gut bacteria. Our vagus nerve will send a message to our brains indicating that our digestive system requires nourishment to keep our microbiome alive (Novik & Savich, 2020).

Some strains will request more food to starve their competitors due to their dietary preferences. Some people can make you want to eat their favorite foods just by looking at them. Maybe you were tricked into eating the cookie. Over the last few generations, there has been a significant shift in our eating habits. Certain bacteria have developed a sweet tooth due to the high sugar and low fiber content of many of the foods we eat these days. However, there is still hope! According to research, changing our diet can alter our gut flora within seven days. There are other indications as well:

- Respiratory system infections (like the common cold).
- Periodontitis and gingivitis
- Candidiasis and yeast infections in the cervix
- Increased triglyceride levels in the blood

9.3 DURATION IT TAKES FOR A PROBIOTIC TO WORK

Figure 9.9: Probiotics and time relations.

Probiotic therapy may have an immediate positive effect on minor health issues, but this is not always guaranteed. A person suffering from mild diarrhea may feel better in a few days. For a few weeks, regular use may not yield digestive health benefits. Because it takes time for the body to develop these beneficial bacteria, taking probiotics regularly may be beneficial. Probiotic effects may take longer or shorter to take effect depending on a variety of factors. A probiotic must have several key components to be successful, including the right strain and dosage and a good formulation and high-quality ingredients. Probiotics are sometimes used as a short-term treatment for acute symptoms like diarrhea caused by travel. A few days later, one may notice the consequences of such an occurrence (Nichols, 2007). Prior research, for example, Reliable Source, claims that, diarrhea can be reduced in duration and frequency by two days by combining probiotics and rehydration. Probiotics should be taken for a longer period for more severe illnesses, dysbiosis, or a weakened immune system.

According to a reputable source review published in Nutrients, probiotic use for at least eight weeks may be the most beneficial for IBS patients (IBS). According to a 2018 study, there was a significant reduction in upper respiratory infections among study participants who drank a high-dose probiotic beverage for 12 weeks. Regular use of probiotic medication has the best chance of relieving symptoms and improving gut health. The number of viable strains in a given probiotic dose is measured in colony-forming units (CFUs). The type of microbe strain, health status, product formula, dosage, and product quality are all critical for effectiveness. The probiotic's action and the timing of the results may differ depending on the condition or symptom that is being attempted to treat. Give it some time to work if you're taking a probiotic to improve your immune system or gut health. However, if you've just had diarrhea, taking a probiotic can help you recover faster.

For example, according to 2010 Trusted Source research, when combined with rehydration therapy, probiotic therapy can reduce the duration and frequency of infectious diarrhea in as little as two days. When compared to a placebo, researchers discovered that taking high doses of *Lactobacillus paracasei, casei,* and fermentium probiotics for 12 weeks significantly reduced upper respiratory infections and flu-like symptoms. It was also discovered that the probiotic beverage improved participants' immune systems after 12 weeks of use by increasing the concentration of antibodies in the stomach (especially sIgA). An additional 2011 Trusted Source study discovered that IBS patients who took *Saccharomyces boulardii* supplementation for four weeks experienced significant reductions in IBS-related symptoms compared to a control group (Narang et al., 2011). Depending on why probiotics are being taken, an improvement in your symptoms may be noticed within a few days to several months.

9.3.1 What could be causing your probiotic to be less effective or to take more time for effect?

Probiotics are not for everyone. Probiotics affect differently depending on genetics, age, health, and the microorganisms in the body.

Is there a reason why probiotics do not always work?

- Using an insufficient amount of the active ingredient (too few CFU).
- You're not doing it right when it comes to eating and fasting (eating vs. fasting) (with food versus on an empty stomach). Reading the product label will help determine how much to take.
- A lack of zeal. Certain strains are better suited to specific ailments.
- The products (live cultures) are of poor quality (live cultures). Brittleness can be caused by probiotics, which is a serious issue. To be effective in the intestines, they must withstand the

manufacturing process, storage, and stomach acid (Ayichew *et al.*, 2017).

- There was an issue with the way they were stored. Probiotics may be harmed by high temperatures, bright light, and humidity. Some people may require some warming up.

9.4 SELECTING THE BEST PROBIOTICS

The intended use determines the best probiotic. Efficacy varies with the situation and level of stress. A wide range of probiotic strains can be found in foods such as yogurt and dietary supplements (Nami *et al.*, 2015). Despite the various claims made by probiotic products, there is compelling evidence that specific probiotics such as Lactobacillus, Bifidobacterium, and *Saccharomyces boulardii* are generally safe and beneficial in some circumstances.

Probiotics may be one of the most promising treatments for many illnesses.

- Travel-related diarrhea prevention and treatment
- Irritable Bowel Syndrome (IBS)
- Antibiotics and diarrhea
- Crohn's disease
- Dry skin

Probiotics are widely accepted to help healthy people maintain their digestive, genital, and immune systems.

9.5 UNDERSTANDING THE THREE RS IS ESSENTIAL FOR SUCCESS.

In excellent physical condition. Because probiotics cannot treat all illnesses, it is critical to match the strain to the specific symptom.

The necessary microbes must exert significant pressure. For the best results, choose a condition that has been proven to work with the help of a reputable source, such as *Bifidobacterium longum* or *Lactobacillus acidophilus* (*Bifidobacterium longum* vs. *Lactobacillus acidophilus*, for instance). Before beginning any new supplement regimens, consult with a doctor.

Correct dosing (CFU) is required. This is dependent on the condition or symptoms you are attempting to treat. Gastrointestinal illnesses were more successfully treated at doses of 5 billion CFUs or higher than at lower concentrations (McKenzie *et al.*, 2018). The recommended dosage varies depending on the brand. Because many products contain multiple bacterial strains, read the label carefully. Medication dosages differ between adults and children.

9.6 IS YOUR PROBIOTIC BENEFICIAL?

Following the dosage instructions on the container and selecting a reputable brand are two of the most important things that can be done to ensure that the probiotic is effective. Each brand will include a different set of information depending on the product.

Probiotic manufacturers are constantly striving to improve the performance of their products and increase their potency and survival through various methods, including microencapsulation.

9.7 HOW TO MAXIMIZE THE BENEFITS OF PROBIOTIC SUPPLEMENTS

To be beneficial, a probiotic must meet the following criteria:

- Excellent quality (living culture). Choose a solution that has a track record of success.
- Excellent condition: When it comes to products, read the labels and store them correctly (in the refrigerator or at room temperature) (refrigeration, room temperature, etc.). Temperatures (freezing or otherwise). In terms of (things like) temperature, refrigeration, and so on.
- Completely followed all instructions. Follow the package instructions and only use this product as directed on the label (before meals, before bedtime, etc.) When it is time to eat when it is time to go to bed, and so on. (Before meals, at night, and so on.)
- Able to live inside a human body. Probiotics must be acid and bile-resistant to establish themselves in your digestive system (Mahasneh & Abbas, 2010).
- Everything will be fine. You have been looked after. Any additional ingredients listed on the label must be considered. Keep an eye out for any fillers or binding agents that may cause issues.

A probiotic label will typically include the name of the probiotic (for example, Lactobacillus acidophilus), the dosage in CFU, the expiration date, and instructions on how to use and store the product. In addition to the expiration date, a "use by" date that specifies how long the product contains viable cultures must be provided. An expiration date based on "time of manufacture" is unreliable. As a result, the cultures may not be present or are not as advertised when you purchase them.

Probiotics have been shown to benefit many medical conditions, including digestive issues such as IBS, bloating, and even Crohn's disease. However, symptoms that are common to many people are rarely studied. As a result, we know that probiotic supplements work, but we do not know how quickly they affect results.

Probiotics have been shown to help alleviate symptoms of irritable bowel syndrome (IBS). Probiotics can help treat IBS and other medical conditions by addressing the underlying causes of symptoms.

- Probiotics can help relieve the symptoms of acute diarrhea for a few days.

Studies show that probiotics treat acute diarrhea in just a few days. Because of the placebo effect, some people will experience amazing results faster than others. People benefit from probiotics because they believe in their efficacy and expect them to be available. Our thoughts influence our lives most (Bowe *et al.*, 2014).

- Probiotics may help people with food intolerances improve over time.

According to preliminary research, probiotics may help with lactose intolerance. A decision can take anywhere from six months to three years to be made.

It is possible that as patients' gut health improves, they will experience fewer symptoms. On the other hand, food intolerances tend to linger for longer periods. Dietary intolerances seem to improve as the gut health improves.

9.8 REACTIONS TO FOOD

Probiotics can be extremely beneficial in identifying and avoiding foods that cause symptoms. Many diets, including the Paleo and low FODMAP diets, can aid in identifying food allergies and improving gut health by identifying food intolerances. Despite the importance of avoiding trigger foods, it is recommended you gradually reintroduce and broaden your diet. Because the path is not always straight, trial and error may be required (Li *et al.,* 2016). It may be necessary to test a food item several times before determining that it is safe to consume a nutrient in moderation.

9.9 METHODS FOR EVALUATING THE EFFECTIVENESS OF YOUR OWN PROBIOTIC

Probiotics, on the other hand, are not a one-size-fits-all treatment for all illnesses. Even if you have previously used probiotics, you should give them a three- to four-week trial run to see if they work for you. You must conduct a thorough experiment and follow a few rules to get the most accurate results.

Choose wisely the first time. People frequently experiment with probiotic supplements without fully comprehending how to use them or which probiotic strains to purchase. Probiotics are a popular dietary supplement, and many people experiment with different brands and strains to see which one works best for them. They may also be concerned about probiotic supplement selection.

Here are some simple guidelines for conducting a successful probiotic experiment. The probiotic high fades after a few weeks.

- Check to see if you are taking a reputable probiotic.

The Food and Drug Administration (FDA) does not evaluate probiotics, but research indicates that some products are not worth purchasing. The two most common reasons for probiotic product failure are:

- For starters, the label promises to be a game changer. Preventing potential health problems.

According to a study finding, only a few, if any, of 26 commercial probiotics tested met their label claims. According to one study, probiotics only contain half of what they claim to contain. According to another study, 43 % of probiotics tested contained less healthy bacteria than advertised. Even the most expensive probiotic supplements are ultimately ineffective (Lin *et al.,* 2019). Avoid probiotic supplements that are significantly less expensive than the alternatives. This is usually an indication of an issue with the quality control procedures.

9.10 THE EFFECTIVENESS OF PROBIOTIC SUPPLEMENTS CAN BE DETERMINED

It is easy to overlook a person's gradual improvement in health. Keep a journal of the symptoms for future reference.

An objective method for assessing the effects of probiotics in the research: First and foremost, pay attention to how you feel.

Fatigue and depression, bloating and joint pain, constipation, insomnia, and an unstable thyroid medication dosage are all possible symptoms. Regardless of the type of symptoms, keep track of how frequently and how severe they are. Check back in 3 to 4 weeks after starting the probiotics.

Please describe the current situation. If this is the case, a definitive answer is not needed to continue the probiotic regimen. If that's the case, that's fantastic. Even so, don't expect to solve all of the problems in weeks. Look for a strong upward trend. If this is the case it is the right track. Continue the excellent work.

The third step is to determine when the progress peaked. Thank you for reaching the pinnacle or highest level of achievement. Furthermore, this is a noteworthy achievement. Continue for another four weeks while the body adjusts to this new level of health (Kumar, 2013).

After four weeks, keep the dosage as low as possible. Although supplements are technically optional, it was discovered that fewer supplements are better. It is recommended that people cut their probiotic supplement dosage in half and reassess it after three to four weeks. If everything goes well, taking probiotics can be discontinued.

There is no one-size-fits-all solution for lowering the probiotic dosage. Consider thinking outside the box. If a recurrence of symptoms is noticed, continue taking probiotics. If necessary, try weaning again in a few months. If you continue to feel better while taking probiotics, you can take them indefinitely.

Symptoms and signs may reappear

When things are going well in life, the digestive health can improve. Regression may occur due to travel, stress, or a prolonged period of poor self-care. Everything will be fine in the end. Take into account the requirements. The Probiotic Triple Therapy regimen can be restarted at a lower dose.

9.10.1 Alternatives to Probiotics in Case They Do not Work

There is no need to try another probiotic supplement after taking a high-quality probiotic supplement and following the Probiotic Triple Therapy protocol for three to four weeks.

The groundwork for future successful gut therapies by using probiotics has been laid. Some people may require more than probiotics to correct long-term bacterial imbalances in the intestines. Herbal antibacterial remedies or even specific antibiotics may be beneficial in this case. According to research, probiotics improve the efficacy of antibacterial treatments (Kukkupuni *et al.,* 2015). Then, if necessary, we will move on to other forms of assistance, such as probiotics.

9.11 WHERE PROBIOTICS WORK IN THE GUT

The normal intestinal microflora is a metabolically active host defense organ that is mostly unknown. The host is subjected to the most severe antigen assault when a normal flora is formed. This antigen challenge significantly increases the development of gut-associated lymphoid tissue. Although bacteria can be found throughout the digestive tract, the large intestine has the most germs and metabolic activity. According to culture-based research, the mouth is home to a varied microbiota that includes facultative and stringent anaerobes such as streptococci, Bacteroides, lactobacilli, and yeasts. The microflora of the upper gut, which comprises the stomach, duodenum, and jejunum, is sparse and can include up to 105 colony-forming units/mL. Bacterial concentrations progressively increase from the ileum to the colon, reaching 1011 to 1012 colony-forming units/g (Kim & Park, 2017). Although at least 500 distinct microbial species are thought to exist, 10-20 genera are expected to predominate in the average human microflora, including *Bacteroides, Lactobacillus, Clostridium, Fusobacterium, Bifidobacterium, Eubacterium, Peptococcus, Peptostreptococcus, Escherichia,* and *Veillonella.*

Microbial colonization starts as soon as the kid is delivered. Bacteria colonizing the infant's gut originate in the vaginal and intestinal flora of the mother. Contact with the environment is another factor in colonization. At this stage, the dominant strains are facultative anaerobes such as lactobacilli, coliforms, and enterobacteria.

Diet may significantly influence the composition and activity of gut bacteria. Breastfed newborns are thought to have a higher concentration of Bifidobacteria than formula-fed newborns, who have a more diverse microbiome that includes enterobacteria, lactobacilli, Bacteroides, clostridia, bifidobacteria, and streptococci. After weaning, the microbiota gradually changes to resemble that of an adult.

On the other hand, the vast majority of naturally occurring bacteria are members of the commensal flora. They have been classified as potentially hazardous or beneficial microorganisms on occasion. The most prevalent strains with these characteristics are Lactobacilli and Bifidobacteria. Bifidobacteria and Lactobacilli are the most common probiotics, and some of these have anti-inflammatory properties. Furthermore, the same genera have been linked to other beneficial qualities like immune response activation and pathogen competitive exclusion, both of which increase non-specific human resistance to microbial diseases (Khoruts, 2018).

9.12 JUSTIFICATION FOR MODIFYING THERAPEUTIC GUT MICROECOLOGY IN INFLAMMATORY AND INFECTIOUS BOWEL DISEASES

The gastrointestinal barrier regulates antigen trafficking in the gut. Numerous components of the intestinal lumen and mucosa influence gut mucosal protection. These components work together to prevent pathogen colonization, remove foreign antigens that have penetrated the mucosa, and control antigen-specific immune responses. The antigen must get past this amazing network of gut defenses to cause infection or inflammation.

When antigens come into contact with the mucosal surface, most of them have already been digested. Proteases in gut bacteria degrade antigenic structure, which is an important step in developing resistance to ingested antigens. The lymphoid compartments of the gut, which include the mesenteric lymph nodes, Peyer's patches, isolated lymph follicles, isolated T cells in the epithelium and lamina propria, and secretory sites, are where the regulatory processes in the intestinal immune response take place (Jugan *et al.*, 2017). Secretory IgA antibodies are located in the gut's common mucosal immune system, which includes the respiratory tract, lacrimal, salivary, and mammary glands. As a result, immune responses generated in lymphoid tissue associated with the gut may influence immune responses at other mucosal surfaces.

Oral tolerance, a distinguishing feature of gut immune control, is described as immunological hyporesponsiveness to antigens received through the intestinal channel. Oral tolerance mechanisms are thought to apply to both local bacteria and food antigens. Clonal deletion and/or energy in the periphery, and a decrease in lymphocyte responsiveness, all contribute to tolerance. T helper 3 cells and regulatory T lymphocytes have been identified as two types of regulatory lymphocytes. These suppressive cytokines are produced, including interleukin 10 and, in a different manner, transforming growth factors. Local cytokine control has been proposed as one of the fundamental mechanisms by which lymphoid tissue associated with the gut maintains homeostasis (Johnston *et al*., 2012). Through oral tolerance to dietary antigens and positive interactions between the indigenous gut bacteria and the epithelium, this homeostasis supports the preservation of tolerance to the considerable antigen exposure offered via the enteric channel. Homoeostasis necessitates proper gut barrier function and epithelial integrity.

Antigen transit rises in the absence of intestinal microflora, indicating that the gut microbiota plays an important role in the intestine's defense barrier.

The initial compositional development of the gut microflora is thought to have a critical role in creating a healthy gut barrier. Early in life, the gut microbiota influences the development of gut-associated lymphoid tissue, which influences systemic and local immune responsiveness, including hyporesponsiveness to antigens derived from bacteria and food. Experiment mice lacking transforming growth factor or interleukin 10 develop an inflammatory response on the mucosa in response to the local gut flora. These findings highlight the complex connections between the gut mucosal barrier and microorganisms.

Intestinal microflora's significance in forming of oral tolerance was investigated in germ-free mice. In contrast, to control mice, germ-free animals were found to retain vulnerability to a systemic immune response, such as the generation of IgE antibodies, after oral antigen administration. The absence of gut flora caused the lack of oral tolerance. Restoring the microbiota at the newborn stage, but not later in life, may be able to correct the abnormal IgE response. A new study compared the culturable microbiota of children born vaginally to those born by cesarean surgery discovered significant differences (Hill, 2010). Colonization was linked to the formation of humoral immune systems, notably circulating IgA and IgM-secreting cells, highlighting the function of gut microbiota in mucosal immune system control.

The usual host-microbe interaction is disturbed in several gut-related inflammatory disorders, and inflammation is accompanied by an imbalance in the intestinal microflora, which may allow local bacteria to provoke an immune response. Rotavirus diarrhca is associated with an increase in fecal urease, an inflammatory mediator that predisposes the gut mucosa to the adverse effects of ammonia and an increase in urease-producing bacteria. According to the findings of Duchmann and colleagues, healthy people can tolerate their microbiota, but inflammatory bowel disease patients lose this ability. Patients with rheumatoid arthritis and allergic disorders have altered gut microbiota, implying that normal gut microflora functions as an ecosystem that responds to inflammation in the gut and other parts of the human body.

The development of gut-associated inflammatory illnesses such as allergies and autoimmune and inflammatory diseases is dependent on the gut microbiota and the dysregulated immune response to their antigenic components, according to early findings from the human study. Probiotic therapy is based on normalizing the characteristics of an unbalanced native microflora by selected strains of healthy gut microflora. Oral probiotics may help stop the inflammation cycle (Harzallah & Belhadj, 2013). The efficacy of the probiotic is demonstrated by a normalization of the altered gut microbiology and increased intestinal permeability and an improvement in the intestine's immunological barrier functions and a reduction in the inflammatory response. Probiotic therapy treats clinical conditions characterized by compromised mucosal barrier function, particularly viral and inflammatory diseases.

CHAPTER 10

POSSIBLE SIDE EFFECTS OF PROBIOTICS

CONTENTS

Probiotics, although helpful microorganisms, may cause unpleasant effects in persons who consume them. Numerous meals include probiotics that the body produces, but if the body is exposed to an excessive amount of these compounds that it is unfamiliar with, it may react unexpectedly and unpleasantly.

There are many dependable and effective probiotic medications on the market. Probiotics' benefits include better digestion, higher nutritional absorption, and boosted immunity.

Probiotics may cause undesirable side effects in a tiny percentage of people. People with low immune systems or those who are unwell may be more vulnerable to the consequences. The effects include;

10.1 BLOATING

A fingerprint has as many variants as there are people. Probiotics are a method of supplementing the body's supply of helpful microorganisms. As a result, Herxheimer Reaction, a short-term sickness, may occur. Trillions of beneficial microorganisms colonize the digestive system. Their presence has an impact on both the exterior and internal gastrointestinal environments. Many hazardous bacteria die because they cannot survive in a favorable environment (Gwee *et al.*, 2018). The vast volumes of chemicals produced by the demise of harmful bacteria are too much for the body to handle. One of the earliest negative effects of taking probiotics is bloating. However, probiotics should be continued to be taken. Drinking plenty of warm water might help relieve bloating.

10.2 DIARRHEA

Probiotics can assist with persistent diarrhea. It is noticed that the stools are a touch watery at first. This is one of the most typical negative effects of taking many probiotics. To avoid this effect, empty-handed, take vitamins first thing in the morning. Shortly after breakfast, the human digestive system is at its busiest. The gut bacteria work particularly hard in the morning to prepare the body ready for the day. You can ensure that your body is filled with beneficial microorganisms first thing in the morning. Raw cheese and yogurt, which can be had in the morning, contain probiotics (Gogineni *et al.*, 2013). If persistent diarrhea is experienced, you may question if probiotics damage your health. The answer is no; however, if diarrhea persists, reevaluate your dose.

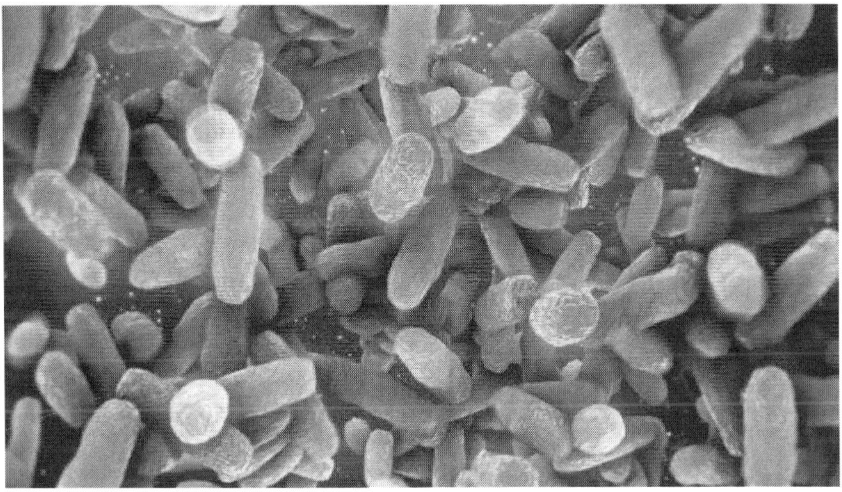

Figure 10.1: The lactobacillus bacteria.

10.3 EXTRA DIGESTIVE PROBLEMS

Frequent mild stomach distention or flatulence usually goes away on its own. Those with compromised digestive systems are more likely to have severe digestive diseases such as diarrhea, constipation, and excessive gas. Mild discomfort is usually in the first few days after starting a probiotic regimen; however, if the symptoms persist, visit a doctor and consider discontinuing probiotics until a more appropriate regimen can be created. Probiotics produced by yeast may cause thirst and constipation in newcomers. The source of probiotics' strong negative effects on certain people is unknown. The best method is to start with a low dose and gradually increase it over a few weeks to the recommended amount.

10.4 HEADACHES

Another common negative effect of probiotics is headaches. Biological amines can be found in various foods, including yogurt, sauerkraut, and kimchi. Biogenic amines are produced when protein-containing foods are aged or fermented by microorganisms. Histamine, tyramine, phenylethylamine, and tryptamine are some of the most common amines in probiotic meals. Amines instantly impact the brain and nervous system (Gupta, 2011). They can cause headaches by stimulating the nervous system and altering blood flow.

In one study, low-histamine meals relieved headaches in 75% of the patients. A separate study found that dietary amines had no discernible effect on headaches based on eleven controlled trials. Additional research is required to understand the effects of dietary amines fully. It is recommended that a dietary log is kept if there is a history of drug sensitivity. This may assist you in determining whether probiotics are safe to consume. You might start with probiotic supplements as an alternative to the complete procedure. Amine levels can only be increased by eating probiotic-rich foods.

10.5 ALLERGIES

Probiotics can cause a variety of allergies. If you are allergic or sensitive to dietary supplements, check the labels carefully. Many dietary supplements, for example, include dairy, soy, and/or egg components. These substances have the potential to cause a severe allergic response. Consult a doctor before beginning a new supplement regimen. If it is discovered that you are allergic to a dietary supplement, you must discontinue use immediately to avoid serious health consequences (Gao *et al.*, 2022). Probiotics can be produced by yeast and bacteria. If you are prone to yeast allergies, take probiotics made from bacteria. Lactose, often known as milk sugar, is included in most dietary supplements.

According to one study, most lactose-intolerant people can take 400 mg of lactose in supplements and pharmaceuticals. However, negative reactions have occurred on occasion. These supplements should be avoided at all costs. There are lactose-free probiotic dietary supplements available. Prebiotic plant fibers feed the good bacteria in the stomach. Humans, on the other hand, are incapable of assimilation (Faujdar *et al.*, 2016). Prebiotics, which encourage the growth of beneficial bacteria colonies in the stomach, are found in various dietary supplements. These drugs are known as synbiotics. Some people may be unable to tolerate prebiotics, resulting in gas or bloating.

It is crucial to bear in mind, though, that not everyone will benefit from every probiotic in the same manner. Some physicians recommend mixing probiotic pills with antibiotics to address a specific condition. As a result, both healthy and hazardous germs could be eliminated, which might create diarrhea. Probiotics help in maintaining or restoring a healthy microbiome in the gastrointestinal system.

10.6 CONSEQUENCES OF HISTAMINE DEFICIENCY

Probiotics may have extra negative consequences if the body perceives the new microorganisms as a threat. Histamine is created by the body when the immune system detects a threat. Certain probiotic bacteria can cause histamine production in the digestive tract. Histamine is naturally degraded in the gastrointestinal tract by the enzyme diamine oxidase (DAO). Histamine intolerance, on the other hand, may cause allergy symptoms in certain people (Ezema, 2013) because their bodies may not produce enough DAO to break down histamine properly. It has the potential to reach the bloodstream via the digestive tract and cause an unpleasant response.

10.7 THE DANGERS OF INFLUENZA

Even though most people believe probiotics are safe, infections have occurred. This is a rare but deadly probiotic adverse effect. Probiotic bacteria or yeast in the bloodstream spread disease. Those with weakened immune systems, intravenous catheters, recent surgery, immunosuppression, or extended hospital stays are more likely to get an infection. Despite this, no major infections have been reported, and the risk of infection is exceedingly low. One in every 5.6 million and one in every one million persons who take Lactobacilli-containing probiotics are at risk of falling sick.

Probiotics are generally safe for most people, and you may already be taking a large dosage of them. However, some people may develop infections, allergies, or digestive issues due to probiotics. The probiotic medication should be stopped if the symptoms last longer than a few days or weeks. Consult a doctor if the side effects of probiotics worsen (Deng *et al*., 2020).

10.8 AMINES IN PROBIOTIC FOODS MAY TRIGGER HEADACHES

10.8.1 What are Amines?

Amino acids are the building blocks of amines, a chemical family. When proteins in the diet are broken down, amines are produced. Beef, for example, does not inherently contain amines, but amines are formed as proteins degrade. Amines can also be produced by overcooked meat. Amines are also created during meal fermentation by the breakdown of proteins. Amines can be present in kombucha, which is a fermented beverage.

Amines can be found in a variety of foods; thus someone sensitive to one type of amine may not be impacted by another. Histamine, a well-known amine, is found in many foods. In addition to histamine, other well-known amines that induce symptoms include serotonin, dopamine, and noradrenaline (Deshpande *et al.*, 2011). Tryptamine and phenylethylamine are two lesser-known amines in the diet. However, amines, such as phenylethylamine in chocolate, may be more frequent. Amines are supposed to serve some purpose.

Unless present in exceptionally high concentrations, amines in food have no negative influence on most people (for instance, "scombroid" poisoning in fish). Some people are so sensitive to amines that even trace amounts in food might trigger symptoms. The phrase "biogenic amines" refers to amines that are present in the body and affect it. They can be absorbed there and participate in several chemical events throughout the body. Biogenic amines have the potential to serve as neurotransmitters, contribute to local immunological responses (such as histamine-induced inflammation), and control gastrointestinal activity. Serotonin, dopamine, and noradrenaline are three important neurotransmitters for brain health (NRA).

- Imbalances can cause anxiety and depression. In addition to symptoms in other body parts, such as headaches, migraines, asthma, and skin rashes, the stomach can also have diarrhea, nausea, and IBS. There is no evidence that amines, like many other dietary compounds, cause migraines. According to experience, salicylates, glutamates, and food additives can cause headaches and migraines in those sensitive to dietary chemicals (Di Cerbo & Palmieri, 2015). According to some (although limited) research, a diet lacking in amino acids may also aid in treating eczema and urticaria.

It is unclear how this may happen. Amine oxidases, such as monoamine oxidases and diamine oxidases (DAO), are typically involved in the breakdown of amines. According to some beliefs, those more sensitive to biogenic amines have a slower rate of enzymatic amine breakdown due to a genetic predisposition or digestive difficulties.

10.9 WHAT FOODS MAY CONTAIN AMINES?

Food has a diverse spectrum of amino acids, making it impossible to forecast how much an individual's diet contains on average. The amino acid content of food is affected by how it is prepared and cooked (for example, how long a piece of meat is browned), if fermentation occurs, the degree of acidity in the food or meal, and storage before cooking. Amines can stimulate the central nervous system, increase or decrease blood flow, and cause headaches in susceptible individuals.

More than 75% of participants who tried a low-histamine diet reported that it helped them with their headaches. A meta-analysis of ten well-conducted trials revealed no detectable effect on headaches (Daliri & Lee, 2015). More study is required to understand whether amines can trigger headaches or migraines in certain persons. Keeping a meal journal will assist in determining whether fermented foods are causing headaches.

If probiotic-rich foods aggravate symptoms, one may benefit more from a probiotic supplement. Fermented foods, which are high in beneficial microorganisms, contain amines naturally. People who have headaches after consuming these foods should take probiotics. Never start taking probiotics without first visiting a doctor or pharmacist. A doctor must prescribe probiotics for those with compromised immune systems or those undergoing cancer treatment.

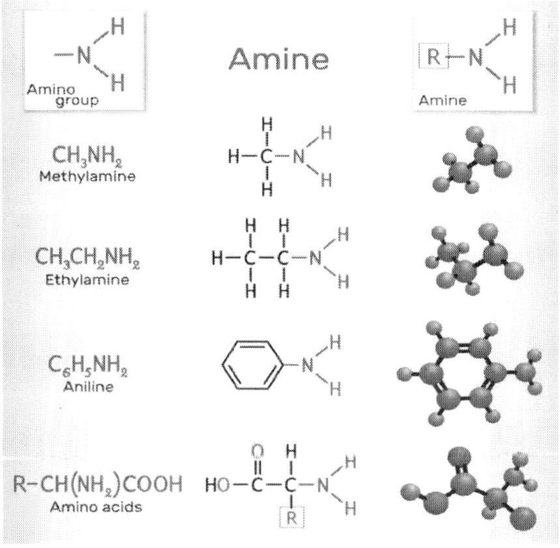

Figure 10.2: Molecules of amines.

The most prevalent probiotic microorganisms are Lactobacillus and Bifidobacterium. When the supplement is taken, these freeze-dried microorganisms become active in the digestive tract. Probiotic supplements can be found at pharmacies and supermarkets. They are available as a supplement in pills, capsules, and a powder that may be mixed into food. It is best to look for a product with a "sell-by" date. Because the amounts of each element vary, it is difficult to provide general dosage guidance. Every day, humans consume between 5 and 10 billion colony-forming units (Charalampopoulos & Rastall, 2009). Each day, only one dosage of probiotics should be taken. Some people may experience loose stools after taking probiotics for a few days, although this usually resolves on its own. Probiotics may help relieve discomfort after a meal.

10.10 EXAMINE THE AMINE SENSITIVITY

People frequently eliminate foods high in certain amines from their diets to test for amine sensitivity. Salades, glutamates, and other dietary elements may have to be eliminated if the RPA's whole elimination diet is followed. A challenging process is used to determine if migraines are caused by amines, salicylates, glutamates, food additives, or a mixture of these compounds. Non-trigger foods are then introduced into the diet, while trigger foods are progressively reintroduced when the patient is ready.

10.11 HISTAMINE-LOWERING DIETS VS. AMINE/ RPAH-ELIMINATION DIETS

If you have headaches or migraines, it makes sense to lower the levels of all amines, not only histamine. Histamine's negative influence on the digestive system might include vomiting, constipation, gas, and stomach discomfort. Another indication of a heart attack is palpitation or a headache, as well as low blood pressure (hypotension). Allergic responses include hives, hay fever, eczema, and asthma (Dai et al., 2018).

Tyramine has been linked to Parkinson's disease, depression, hypertension, and migraines. This chemical has been connected to schizophrenia and attention-deficit/hyperactivity disorder. Histamine diet programs are accessible online and in a variety of formats.

10.11.1 Items to Avoid While Following a Low-Histamine Diet

The meal creates histamine when consumed; however, it is naturally low in histamine.

When lists are combined, these two distinct list types might pose a challenge.

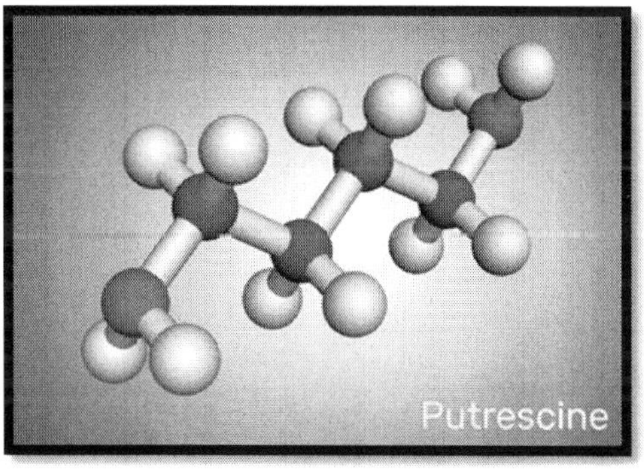

Figure 10.3: Putrescine molecule.

A crucial component of the RPAH elimination diet is excluding foods rich in histamine and amines. A diet high in foods low in histamine has a broader influence. Because it is expected to be more effective, salicylates, glutamates, and other dietary additives are avoided as part of the RPAH elimination package. In contrast, the RPAH elimination diet removes foods that are low in amines yet are considered to increase histamine synthesis in the body (Cammarota *et al.*, 2014). If it is believed that the food is causing headaches or migraines, dietary amines should be investigated. If there is a link, eliminating the trigger should improve your symptoms; setting up routines to track your symptoms might help.

10.12 SOME INGREDIENTS MAY CAUSE ADVERSE REACTIONS

Food additives such as flavor enhancers, preservatives, and coloring pigments may cause allergic reactions in people with food allergies. Rashes, diarrhea, other gastrointestinal disorders, and respiratory illnesses such as asthma are all symptoms of food additive allergies. According to the Asthma & Allergy Foundation of America, eight additives may cause allergic responses. Sulfonamides Monosodium glutamate, or MSG for short, is a flavor enhancer. An allergic response to this ubiquitous Asian spice can include asthma attacks, chest tightness and back of the throat tightness, nausea, vomiting, diarrhea, and a burning feeling in the upper chest and neck.

10.12.1 Yellow Hues

Yellow dyes Nos. 5 and 6 (tartrazine 102 and yellow 2G107) are used to color various foods (sunset yellow FCF 110). (FCF 110 Sunset Yellow) (FCF 110 Yellow of Sunset) The human body produces benzidine, a well-documented carcinogen, from other yellow-colored compounds. Yellow No. 1 dye Children allergic to aspirin may exhibit symptoms comparable to an allergic reaction (Chugh & Kamal-Eldin, 2020). Yellow dye No. 6 has been linked to severe hypersensitivity reactions in certain persons and instances of adrenal and kidney cancer. Yellow coloring may be found in a variety of meals.

- It can produce allergic-like symptoms in aspirin hypersensitive people and hyperactivity in certain youngsters.

10.12.2 Sulfites

Sulfites were outlawed by the FDA in 1986 as a technique to maintain the color and freshness of fresh fruits and vegetables. Sulfite allergies can cause itching, skin rashes, and nausea. Asthma medicines containing sulfite are also widely accessible. 5% of asthmatics, on the other hand, may have anaphylactic responses to sulfites, which include:

- Chest dissatisfaction
- Abdominal aches and pains
- Diarrhoea and nausea
- Breathing difficulties

This strong reaction might be explained by sulfur dioxide hypersensitivity. The FDA banned using sulfites to preserve the color and flavor of fresh fruits and vegetables in 1986. People sensitive to sulfite are more likely to have symptoms such as itching, skin rashes, and gastrointestinal discomfort (Choudhary & Tandon, 2009).

- BHT and BHA both

Butylated hydroxytoluene (BHT) and butylated hydroxyanisole (BHA) are common ingredients in breakfast cereals, loaves of bread, and other grain-based products to help prevent oxygen-induced grain deterioration while also keeping color, taste, and fragrance. Those allergic to BHT and BHA may have skin edema, redness, hives, and extreme itching.

10.12.3 Sodium Bromate Potassium Bromate

Potassium bromate, a chemical compound utilized in the making of baked products, is present in a variety of them. According to the WHO's International Agency for Cancer Research, this additive has caused animal cancer. After baking, potassium bromate, a toxin, remains in the bread. Besides Japan and the United States, bromate is prohibited in all nations. Potassium bromate is used in the manufacture of a range of baked goods.

10.12.4 Substances of Various Kinds

Other ingredients, such as aspartame, nitrates, nitrites, benzoates, and others, might cause allergic reactions. As far as we know, these additives do not produce many side effects. In other circumstances, such as the calorie-free sweetener aspartame, which is used to sweeten a wide range of foods and beverages, symptoms such as hives and dilated pupils have not been seen. Although nitrates and nitrites are food preservatives, flavor enhancers, and botulism inhibitors, excessive consumption can cause headaches and skin rashes (Chalas *et al.*, 2016). There have been minimal reactions to benzoates, widely employed as preservatives in baked goods and other foods.

Aspartame, nitrates, nitrites, and benzoates are other substances that may cause allergic reactions. Although nitrates and nitrites are used to preserve, enhance, and prevent botulism, they might cause a headache or rash.

- Foods Containing Sulfite

Sulfites, inorganic salts, are widely utilized as preservatives in a wide range of foods. Sulfites, among other things, hinder the growth of germs, prevent food browning, and bleach some starches and cherries. Some persons can acquire sulfite sensitivity. Asthmatics are more prone to develop an allergic reaction to sulfite-containing meals. Since 1988, four government agencies have monitored sulfites in medications, food, meat and poultry, and alcoholic beverages. Knowing if foods contain sulfites helps recognize and prevent sulfite sensitivities (Culligan *et al.*, 2009).

10.13 HOW DO INFECTIONS OCCUR?

A person becomes infected when bacteria enter the body, reproduce, and activate the immune system. An infection can manifest itself in three ways:

- Harmful microorganisms (germs) on human skin and other surfaces are the source (e.g., sinks, surfaces, human skin)
- A person who is susceptible to infection due to a germ entry site.
- Pathogens can be conveyed to a vulnerable individual via transmission.
- While most people can use probiotics safely, not everyone will benefit from them.

Probiotic bacteria or yeasts can enter the bloodstream and infect susceptible humans in rare situations. Individuals who have recently had surgery, those who have venous catheters, and those who have been in the hospital for a lengthy period are all more likely to get an infection. Infections. Although infections are uncommon, population-based clinical investigations have shown no indication of a relationship between the condition and serious infections (Culligan *et al.,* 2009). According to estimates, one in a million persons who take probiotics containing Lactobacilli bacteria may fall sick. The 5,600,000 people taking yeast-based probiotics have a far reduced risk of illness than the overall population.

Infections are possible, but most are curable with antibiotics or antifungal drugs. However, there have been a few fatalities. Individuals with severe acute pancreatitis should avoid taking probiotics for these and other reasons. Probiotics should not be used if the immune system is impaired, you have intravenous catheters, you have just had surgery, you have acute pancreatitis, or you have been in the hospital for a long time (Attri *et al.*, 2021). Consuming a significant quantity of live microorganisms known as "probiotics" can provide health advantages. These probiotics are readily available and can be obtained in fermented foods or as nutritional supplements.

- While some people may have negative responses to probiotics, the vast majority do not. The most prevalent transient side effects are gas, bloating, constipation, and thirst. Certain persons may experience adverse responses to supplements and probiotic foods containing naturally occurring amines. Probiotics should no longer be utilized (Prins, 2011). Those with compromised immune systems, who have spent time in the hospital or have undergone surgery are less prone to infections caused by probiotic bacteria. Before determining whether or not to use probiotics, weigh the risks and benefits. When taken in a balanced diet or supplement regimen, probiotics are typically well accepted and have few harmful side effects.

10.14 THEY MAY CAUSE UNPLEASANT DIGESTIVE SYMPTOMS

When taken orally, probiotics, which are living bacteria, have been demonstrated to benefit health. Purely Inspired Probiotics may help with weight loss. They have been found in fermented foods and supplements. Probiotics provide several health advantages, including an improved immune, digestive, and cardiovascular health (Qi et al., 2019). According to several research studies, probiotics may also aid in weight loss and the reduction of stubborn belly fat. The microbiota in the stomach may be associated with body weight policy. A large range of bacteria lives in the gastrointestinal system.

The vast majority of these microbes create essential minerals such as vitamin K and a variety of B vitamins. Furthermore, they improve the gastrointestinal system's conversion of indigestible fiber into useful short-chain fatty acids like butyrate.

The two most frequent forms of large gut bacteria are Bacteroides and firmicutes. Bacterial diversity appears to be inversely related to body weight in these two groups. According to human and animal studies, persons who are overweight but not obese have a distinct gut flora than those who are obese (Reid, 2016). In the great majority of these investigations, obese adults had lower firmicute and higher Bacteroides counts than those of normal weight. In all of the studies conducted, the ratio of firmicutes to bacteroidetes has not been associated with obesity.

10.15. SIDE EFFECTS OF PHENQ DIETARY SUPPLEMENTS

The components list of PhenQ has no synthetic or hazardous chemicals. So far, no negative effects have been observed in most patients who take this product for weight loss. A few people have had negative experiences, such as:

- Constipation that lasts for longer than a few days
- Sleepiness
- Stomach dissatisfaction

To avoid unpleasant side effects, take PhenQ as directed daily. If any unfavorable side effects are experienced while taking PhenQ, discontinue use and see a doctor. If you take prescription medications or contraception, see a doctor before commencing the PhenQ program. To avoid overdosing on PhenQ, avoid using other dietary supplements containing the same natural ingredients as PhenQ. They cause stomach distress (Rusu *et al.*, 2019). Even though most people do not suffer any unwanted side effects, probiotic tablets containing bacteria are widely said to produce a transient rise in gas and bloating. As side effects, Candida-based probiotics may induce constipation and increased thirst.

Some people may have adverse effects, but after a few weeks of consistent use, the cause is generally forgotten. To reduce the possibility of side effects, begin with a modest dose of probiotics and gradually increase to the full amount over a few weeks. As a result, the body will be able to adjust. If gas or bloating has been experienced for more than a few weeks, it is advisable to discontinue use. Side Effects of Probiotic Weight Loss Medicine

Bacteria in probiotic supplements may trigger histamine production in the human digestive tract.

The histamine molecule is routinely produced when the immune system perceives a threat. Blood arteries expand when histamine levels rise, allowing more blood to flow to the afflicted area. The arteries become more porous, allowing immune cells to penetrate the affected area faster and eliminate any invaders that may be lurking there (Singh & Natraj, 2021).

Itching, watery eyes, a runny nose, and breathing difficulties are the symptoms. Diamine oxidase is the enzyme responsible for histamine breakdown in the digestive system. Consequently, histamine levels do not grow to the point where they may cause symptoms. Those with histamine tolerance cannot break down histamine in their systems because they do not make enough DAO. The stomach lining excretes excess histamine into the circulation, resulting in allergy-like symptoms.

10.16 SIDE EFFECTS OF PROBIOTIC WEIGHT LOSS MEDICINE

Individuals who are sensitive to histamine should avoid foods high in it. Probiotic capsules that do not include histamine-producing bacteria are the greatest choice for customers, but little study has been conducted in this area thus far.

- Some compounds can cause unpleasant responses.

Those with allergies or intolerances should carefully examine probiotic supplements since they may include allergens or triggers for intolerances.

- Some vitamins may include allergens in dairy, egg, or soy.

Anyone allergic to these ingredients should avoid them since they may trigger an allergic reaction. Make sure to thoroughly read the labels to avoid them. Those who are yeast sensitive should avoid probiotics that include yeast. Instead, probiotics, which contain microorganisms, are required (Singh *et al.,* 2011).

- Lactose is a common ingredient in probiotic supplements.

Despite research suggesting that many people with lactose intolerance may take 400 mg of lactose per day in supplements or pharmaceuticals, there have been adverse reactions to probiotics. Lactose-sensitive persons may have gas and bloating if they ingest lactose-containing probiotics; in this case, they should choose lactose-free products. Certain drugs include prebiotics and powerful probiotics. Plant fibers that bacteria may digest are not safe for humans to consume. Oligosaccharides are classified into three types: lactulose, inulin, and inulin. Synbiotics are food supplements that include both probiotic microorganisms and prebiotic fiber.

10.17 SOME STRAINS CAN INCREASE HISTAMINE LEVELS

Histamine, like dopamine, noradrenaline, and serotonin, is a biogenic amino acid. These critical building blocks may be found in both animal and plant sources of biogenic amino acids. They can also arise in high-protein meals due to processing, aging, and storage. Histamine comes from the Greek word "histos," which translates to "tissue" in English (Sidhu *et al.,* 2015).

Humans require a small quantity of histamine for a range of processes such as breathing, temperature control, diurnal cycles, stomach acid generation, and immunity. Histamine is often kept in basophilic granulocytes and mastocytes (a kind of white blood cell) when generated (type of white blood cell). Other "trigger" molecules, such as ethanol, some medications, and certain meals, can also cause histamine to be produced in response to an allergic reaction or inflammation. The phrase "endogenous histamine" refers to histamine generated naturally by the body.

Histamine may be found in various foods, including chocolate, mushrooms, spinach, and aubergine, as well as meals that have been fermented, matured, or otherwise stored for a lengthy period. Exogenous histamine is histamine that is consumed with food. Our cells create histamine when we consume particular foods, such as strawberries and tomatoes. Histamine releasers suit the bill.

10.17.1 Viruses and Hepatitis C

A diverse range of microorganisms can create histamine. These bacteria may be found in salami (aged cheese), sauerkraut (sauerkraut), red wine (sauerkraut), soy sauce (soy sauce), and beer. Some probiotic bacteria can create histamine.

10.18 WHAT EXACTLY IS "HISTAMINE INTOLERANCE"?

Histamine exposure and breakdown have similar effects on a healthy person; therefore there are no negative health implications. Histamine intolerance patients confront difficulties because their body's normal equilibrium is disrupted, resulting in excess of histamine. The body produces too much histamine or consumes too much histamine-containing food, which can induce this response. When diamine oxidase (DAO) is not functioning correctly, there is an excess of histamine in the body (Arora & Baldi, 2015). DAO breaks down histamine in the digestive tract; when it is inadequate, external histamine accumulates. Genetic anomalies, drugs, and gastrointestinal disorders such as Crohn's disease or ulcerative colitis all contribute to this syndrome.

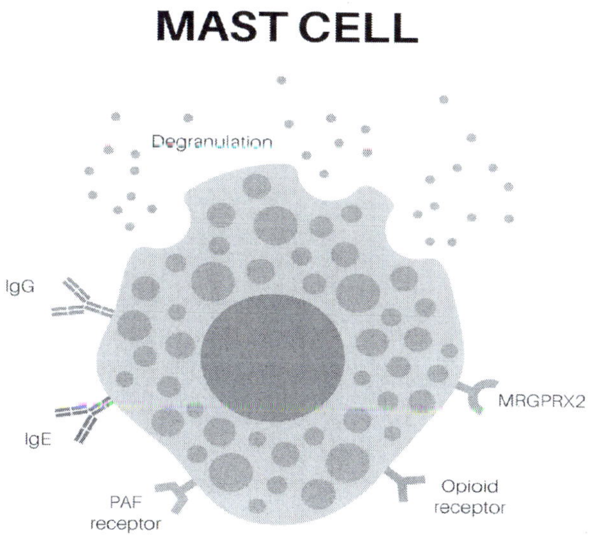

Figure 10.4: Mast cell.

10.19 HOW IS HISTAMINE INTOLERANCE IDENTIFIED?

To avoid histamine intolerance, the amount of time exogenous histamine may be supplied must be limited. This involves avoiding meals with high histamine levels. You may be wondering how probiotics might benefit persons who are histamine allergic. Bacteria have a profound impact on histamine levels in the body. Histamine may be produced in the stomachs of persons who consume excessive levels of this amino acid. Probiotics help the immune system recuperate by aiding in the breakdown of histamine. A particular strain of probiotics can be utilized to cure a particular ailment.

10.19.1 Probiotics with a Reduced Histamine Concentration

If histamine allergy exists, avoid probiotics that increase histamine levels. The health is deteriorating rather than improving due to the probiotic dosage (Aponte et al., 2020). Probiotics that can help with histamine reduction include:

- *Lactobacillus gasseri* is a bacterium.
- Lactic acid-producing bacteria
- *Lactobacillus plantarum* larvae

- sp. tobacillus
- *Lactobacillus reuteri* is the bacterium in this case.
- *Lactobacillus ifidobacterium* sp.
- *Bifidobacterium infantum* is a bacterium.
- *Bidobacterium breve* (*Bifidobacterium breve*) is the scientific name
- *Bifidobacterium longum* is a bacterium.
- *Bifidobacterium bifurcationum*

Probiotics that degrade histamine have a specific effect on both histamine and acidity. *Bifidobacterium longum, Lactobacillus rhamnosus*, and *Bifidobacterium infantis* are suitable alternatives for persons with sensitive stomachs.

10.20 PROBIOTICS THAT CAUSE HISTAMINE RELEASE

Sometimes, probiotics may reverse, raising histamine levels rather than decreasing them. Probiotics that have been proven to increase histamine levels in the human body include:

- *Lactobacillus casei* is a bacteria.
- Lactobacillus larvae from Bulgaria
- *Lactobacillus delbrueckii* is the species.
- SL5 *Bacillus coagulans*

Temperature tolerant microorganisms Thermophilus

Choosing a Low Histamine Probiotic Dietary Supplement

Given the abundance of histamine-degrading probiotics on the market, choosing the best one may be difficult. Before making a purchase, consider the following:

10.20.1 Select the Best Probiotic Strain

Each probiotic pill includes a different strain of bacteria. It is important to understand that the strain of an organism, rather than its genus or species (e.g., Lactobacillus, Bacillus, rhamnosus, or plantarum), is the most important factor (e.g., GG, GR-1). *Lactobacillus rhamnosus*, for example, has a distinct effect on the body. As a result, before purchasing any probiotic supplement, ensure that the label or container details the strain's qualities (Sistek *et al.,* 2006).

10.20.2 Investigate the Benefits and Drawbacks Properly

Before choosing a probiotic, make sure it is the right choice. This suggests investigating any potential negative health impacts. We must address some physical incompatibility. We may develop an allergy or another unwanted reaction if we ingest an incompatible material. Our position may worsen rather than improve as a result of tackling intolerance.

10.20.3 Choose Tried-and-True Brands and Variations

Do not buy something only because of its appealing package or well-known brand name. Rather than relying just on theory, you should investigate if these probiotics have been subjected to human testing. Some probiotics are tested on animals rather than humans. As a result, what works well for animals other than humans may not work well for people (Stavropoulou & Bezirtzoglou, 2020).

10.20.4 Consider Using Single-Strain Probiotics

If the health is really important, probiotics may be taken with 10 to 30 strains. Probiotics with a single strain are preferred to those with many strains; after hypersensitivity has been reduced, switch to a new strain.

10.21 THE CORRECT DOSE

Never take probiotics in random order. Consider the tolerance level. A tenth of the indicated dose is a good starting point. The doses can then be adjusted lower or higher based on your tolerance. If too many probiotics are ingested, the body may respond differently, worsening rather than curing the intolerance (Saraf *et al.,* 2010).

10.21.1 Consider the Accessibility

It may be tough to return brands that are hard to find. In almost all circumstances, after-sales service is critical for maintaining the customer-supplier connection. Choose probiotics that have been scientifically confirmed, well-established, and are well-known. Because of the various ways our bodies are built and structured, we all have differing amounts of histamine. This explains why our systems react differently to meals and probiotics that might boost or lower histamine levels. The essential thing to remember is to become acquainted with our bodies so that we are aware of their requirements and potential risks (Aggarwal *et al.*, 2013). Although not everyone is susceptible to histamines, we do not have to wait until we are affected before taking action. As a result, we must be careful about what we put into our bodies. If you ever have this issue, figure out which probiotics are best for you.

10.22 WHAT YOU SHOULD KNOW ABOUT PROBIOTICS AND HISTAMINE INTOLERANCE

Probiotics are important nutrition for treating histamine sensitivity. Many individuals, however, are unaware that most probiotics create histamine. What exactly is it? That is correct. Even the healthy bacteria you have been taking daily might be exacerbating the symptoms. Consuming low-histamine probiotics (the enzyme that helps break down histamine) is critical for maintaining a healthy gut and increasing DAO enzyme synthesis. I will now go through which probiotic strains (and even brands) are the most effective for histamine intolerance.

10.23 THE IMPORTANCE OF PROBIOTICS IN THE TREATMENT OF HISTAMINE INTOLERANCE

Our bodies contain around 100 trillion bacteria, participate in all human activities, from immunological health to mood and energy levels. When considering the importance of gut flora on symptoms, histamine intolerance is not an exception._Indeed, in the earlier post on the causes of histamine intolerance, it is discussed how gut bacteria might contribute to histamine symptoms and be the primary underlying cause of histamine sensitivity. Bacterial involvement will always be present due to the several systems impacted by histamine intolerance and the disorder's multisymptomatic character (Smolyansky, 2010). Choosing the right probiotic for histamine intolerance is always critical. The most common error people with histamine intolerance make when picking probiotics is selecting the wrong strain, which worsens their symptoms.

10.23.1 Bacteria's Role in Histamine Intolerance

Bacteria are involved in both histamine synthesis and breakdown. Certain types of bacteria in the stomach can convert the semi-essential amino acid histidine into histamine when people ingest meals containing it. Even if the item was not previously deemed histamine-rich, this conversion raises internal histamine levels. Histidine to histamine conversion is frequently harmless and even beneficial.

Histamine intolerance, on the other hand, can result in an accumulation of excess histamine in the body, and an imbalance of gut flora may be responsible for histamine breakdown. A lack of these histamine-degrading bacteria contributes to a persistent rise in histamine levels in the body. Histamine intolerance can be caused by various bacterial processes (Tadesse, 2012). However, why are the gut microbes out of whack in the first place? Disease, pathogen invasion, dietary variations, antibiotic usage, chronic stress, and other environmental and lifestyle factors can all lead to out-of-control bacterial populations.

The generation of histamine by bacteria is the same reason those histamine-sensitive persons should avoid fermented foods (bone broth, kefir, wine, and cheese). Bacteria transform your wines and cheeses into histamine-rich meals in addition to digesting and fermenting them. In almost all cases of histamine intolerance, eating low-histamine meals and choosing the right probiotic is critical.

10.24 RESTORING BACTERIAL BALANCE IN THE GUT

Many people use probiotics for histamine sensitivity to restore gut flora equilibrium. Many studies have demonstrated that probiotics boost gut bacterial populations and general health. As a result, this is a good overall answer. Histamine intolerance exacerbates the issue. Buying probiotics from the store is one of the biggest blunders that can be made if you are histamine sensitive. Whether it has been heard, it is the greatest probiotic on the market (Valero-Cases *et al.*, 2020). Probiotics include virtually universally histamine-producing bacteria, which have the potential to disrupt the bacterial balance and aggravate symptoms. This has happened to many customers, and as can be expected, it is not nice. It is an unneeded waste of time, money, and misery. The objective is to alleviate pain and make this journey less difficult. An extensive research has been done to supply greatest grade probiotics for histamine intolerance. Below is a list of probiotics that have been proved to be good for histamine intolerance, those that should be avoided, and others that are presently being debated (or may only alleviate some symptoms)

- When the number of anti-inflammatory cells grows

The probiotic *Bifidobacterium longum* destroys histamine. It may help reduce the post-meal inflammatory response and prevent or improve intestinal hyperpermeability ("leaky gut syndrome"). This bacterium may aid histamine and tyramine degradation (Abuajah *et al.,* 2015).

- *Lactobacillus plantarum* - May aid in the breakdown of biogenic amines, including histamine and tyramine.

10.25 PROBIOTIC STRAINS TO AVOID

These species can either cause histamine production or contain histamine.

- *Lactobacillus casei* is a bacteria.
- *Lactobacillus bulgaricus* (*Lactobacillus bulgaricus*)
- *Thermococcus thermophilus*
- *Lactobacillus delbrueckii* is the species.
- *Lactobacillus reuteri* (*Lactobacillus reuteri*)

Stresses of Extra Importance

It was discovered that several strains that are either being studied actively or have successfully treated certain conditions.

- *Lactobacillus reuteri* - Although this bacterial strain appears on numerous low-histamine food lists, *Lactobacillus reuteri* is an exception. It not only raises histamine levels but also raises anti-inflammatory cyclic adenosine monophosphate (cAMP) (Wescombe *et al*., 2012).
- *Saccharomyces boulardii* - A histamine-free probiotic that aids in the regulation of digestive illnesses such as diarrhea.
- *Lactobacillus lactis* - Scientists and medical experts are currently debating this strain.
- *Lactococcus lactis* - This strain is used to produce a range of high-histamine foods, yet research has shown that it is histamine-free.
- *Lactobacillus acidophilus* - Medical specialists have extensively researched this strain, and lower doses (less than one billion CFU) appear to be histamine-neutral while also lowering inflammation and improving gut health. Although compiling a list of low-histamine bacteria is simple, finding a probiotic for histamine sensitivity in stores may be tough (Yao et al., 2020). High

histamine species from the 'avoid' or 'questionable' lists, such as Lactobacillus acidophilus (in levels greater than 1B CFU) or *Streptococcus thermophilus*, are among the most often available in stores and pharmacies.

In this comprehensive research on probiotics for those with histamine sensitivity, it was discovered that this is the greatest low-histamine probiotic since it contains histamine-friendly strains solely. The bulk of the clients have no trouble incorporating this probiotic. When combined with a low-histamine diet, this probiotic supports a healthy balance of gut flora and alleviates histamine intolerance symptoms. Life is far too brief to be ruled by symptoms.

The ability of probiotic foods to restore the gut microbiota, which is vital for maintaining the balance between health and disease processes, benefits both firms and customers. The discovery and development of recombinant microorganisms with antiviral properties have advanced in recent years. Looking into probiotics in the context of milk was a key element of this chapter's focus (Zhang *et al.*, 2016). Microorganisms, products, and strategies for developing innovative, risk-free functional foods are also covered. Food items that give extra health benefits in addition to nutrition are becoming increasingly popular as people become more aware of the link between a balanced diet, lifestyle, and overall health.

Maintaining human health and wellness is dependent on the consumption of the right probiotic strains, according to research. A probiotic is a probiotic microorganism. Probiotics, as defined by the FAO/WHO, are "living microorganisms that give a health benefit to the host when fed in suitable amounts." Probiotics have good effects on the body, including the maintenance or stimulation of several physiological processes and the decrease of symptoms and risk factors associated with certain health issues. Probiotics have been linked to many health benefits, including improved gastrointestinal health and modulation of the immune system.

REFERENCES

1. Abuajah, C. I., Ogbonna, A. C., & Osuji, C. M. (2015). Functional components and medicinal properties of food: a review. *Journal of food science and technology*, *52*(5), 2522-2529.
2. Aggarwal, J., Swami, G., & Kumar, M. (2013). Probiotics and their effects on metabolic diseases: an update. *Journal of clinical and diagnostic research: JCDR*, *7*(1), 173.
3. Aponte, M., Murru, N., & Shoukat, M. (2020). Therapeutic, prophylactic, and functional use of probiotics: a current perspective. *Frontiers in Microbiology*, *11*, 562048.
4. Arora, M., & Baldi, A. (2015). Regulatory categories of probiotics across the globe: a review representing existing and recommended categorization. *Indian journal of medical microbiology*, *33*, S2-S10.
5. Attri, S., Singh, N., Nadda, A. K., & Goel, G. (2021). Probiotics and Their Potential Applications: An Introduction. *Advances in Probiotics for Sustainable Food and Medicine*, 1-26.
6. Ayichew, T., Belete, A., Alebachew, T., Tsehaye, H., Berhanu, H., & Minwuyelet, A. (2017). Bacterial probiotics their importances and limitations: a review. *J Nutr Health Sci*, *4*(2), 202.
7. Bowe, W., Patel, N. B., & Logan, A. C. (2014). Acne vulgaris, probiotics and the gut-brain-skin axis: from anecdote to translational medicine. *Beneficial microbes*, *5*(2), 185-199.
8. Charalampopoulos, D., & Rastall, R. A. (Eds.). (2009). *Prebiotics and probiotics science and technology* (Vol. 1, p. 354). New York: Springer.

9. Cammarota, G., Ianiro, G., Bibbo, S., & Gasbarrini, A. (2014). Gut microbiota modulation: probiotics, antibiotics or fecal microbiota transplantation? *Internal and emergency medicine*, *9*(4), 365-373.
10. Culligan, E. P., Hill, C., & Sleator, R. D. (2009). Probiotics and gastrointestinal disease: successes, problems and future prospects. *Gut pathogens*, *1*(1), 1-12.
11. Chalas, R., Janczarek, M., Bachanek, T., Mazur, E., Cieszko-Buk, M., & Szymanska, J. (2016). Characteristics of oral probiotics–a review. *Current Issues in Pharmacy and Medical Sciences*, *29*(1), 8-10.
12. Choudhary, R., & Tandon, R. V. (2009). Consumption of functional food and our health concerns. *Pakistan Journal of Physiology*, *5*(1).
13. Chugh, B., & Kamal-Eldin, A. (2020). Bioactive compounds produced by probiotics in food products. *Current Opinion in Food Science*, *32*, 76-82.
14. Dai, C. E., Li, H. L., He, X. P., Zheng, F. F., Zhu, H. L., Liu, L. F., & Du, W. (2018). Research advance in metabolism of effective ingredients from traditional Chinese medicines by probiotics. *Zhongguo Zhong yao za zhi= Zhongguo Zhongyao Zazhi= China Journal of Chinese Materia Medica*, *43*(1), 31-38.
15. Daliri, E. B. M., & Lee, B. H. (2015). New perspectives on probiotics in health and disease. *Food Science and Human Wellness*, *4*(2), 56-65.
16. Di Cerbo, A., & Palmieri, B. (2015). The market of probiotics. *Pakistan journal of pharmaceutical sciences*, *28*(6).
17. Deshpande, G. C., Rao, S. C., Keil, A. D., & Patole, S. K. (2011). Evidence-based guidelines for use of probiotics in preterm neonates. *BMC medicine*, *9*(1), 1-13.
18. Deng, X., Tian, H., Yang, R., Han, Y., Wei, K., Zheng, C., & Chen, T. (2020). Oral probiotics alleviate intestinal dysbacteriosis for people receiving bowel preparation. *Frontiers in medicine*, *7*, 73.
19. Ezema, C. (2013). Probiotics in animal production: A review. *Journal of Veterinary Medicine and Animal Health*, *5*(11), 308-316.
20. Faujdar, S. S., Mehrishi, P., Bishnoi, S., & Sharma, A. (2016). Role of probiotics in human health and disease: an update. *International Journal of Current Microbiology and Applied Sciences*, *5*(3), 328-344.
21. Gao, J., Wang, R., Liu, J., Wang, W., Chen, Y., & Cai, W. (2022). Effects of novel microecologics combined with traditional Chinese medicine

and probiotics on growth performance and health of broilers. *Poultry science*, *101*(2), 101412.

22. Gupta, G. (2011). Probiotics and periodontal health. *Journal of medicine and life*, *4*(4), 387.
23. Gogineni, V. K., Morrow, L. E., Gregory, P. J., & Malesker, M. A. (2013). Probiotics: history and evolution. *J Anc Dis Prev Rem*, *1*(2), 1-7.
24. Gwee, K. A., Lee, W. W. R., Ling, K. L., Ooi, C. J., Quak, S. H., Dan, Y. Y., ... & Wong, C. Y. (2018). Consensus and contentious statements on the use of probiotics in clinical practice: a south east Asian gastro□neuro motility association working team report. *Journal of gastroenterology and hepatology*, *33*(10), 1707-1716.
25. Harzallah, D., & Belhadj, H. (2013). Lactic acid bacteria as probiotics: characteristics, selection criteria and role in immunomodulation of human GI muccosal barrier. *Lactic Acid Bacteria—R & D for Food, Health and Livestock Purposes; Kongo, M., Ed*, 197-216.
26. Hill, C. (2010). Probiotics and pharmabiotics: alternative medicine or an evidence-based alternative? *Bioengineered bugs*, *1*(2), 79-84.
27. Johnston, B. C., Ma, S. S., Goldenberg, J. Z., Thorlund, K., Vandvik, P. O., Loeb, M., & Guyatt, G. H. (2012). Probiotics for the prevention of Clostridium difficile–associated diarrhea: a systematic review and meta-analysis. *Annals of internal medicine*, *157*(12), 878-888.
28. Jugan, M. C., Rudinsky, A. J., Parker, V. J., & Gilor, C. (2017). Use of probiotics in small animal veterinary medicine. *Journal of the American Veterinary Medical Association*, *250*(5), 519-528.
29. Khoruts, A. (2018). Targeting the microbiome: from probiotics to fecal microbiota transplantation. *Genome medicine*, *10*(1), 1-3.
30. Kim, J. M., & Park, Y. J. (2017). Probiotics in the prevention and treatment of postmenopausal vaginal infections. *Journal of Menopausal Medicine*, *23*(3), 139-145.
31. Kukkupuni, S. K., Shashikumar, A., & Venkatasubramanian, P. (2015). Fermented milk products: probiotics of ayurveda. *Journal of Medical Nutrition and Nutraceuticals*, *4*(1), 14.
32. Kumar, A. (2013). Probiotics: Nature's medicine. *International Journal of Nutrition, Pharmacology, Neurological Diseases*, *3*(3), 219.

33. Lin, T. L., Shu, C. C., Lai, W. F., Tzeng, C. M., Lai, H. C., & Lu, C. C. (2019). Investiture of next generation probiotics on amelioration of diseases–Strains do matter. *Medicine in Microecology*, *1*, 100002.
34. Li, C., Li, X., Han, H., Cui, H., Peng, M., Wang, G., & Wang, Z. (2016). Effect of probiotics on metabolic profiles in type 2 diabetes mellitus: A meta-analysis of randomized, controlled trials. *Medicine*, *95*(26).
35. Mahasneh, A. M., & Abbas, M. M. (2010). Probiotics and traditional fermented foods: The eternal connection (Mini-Review). *Jordan Journal of Biological Sciences*, *3*(4), 133-140.
36. McKenzie, V. J., Kueneman, J. G., & Harris, R. N. (2018). Probiotics as a tool for disease mitigation in wildlife: insights from food production and medicine. *Annals of the New York Academy of Sciences*, *1429*(1), 18-30.
37. Nami, Y., Haghshenas, B., Abdullah, N., Barzegari, A., Radiah, D., Rosli, R., & Khosroushahi, A. Y. (2015). Probiotics or antibiotics: future challenges in medicine. *Journal of medical microbiology*, *64*(2), 137-146.
38. Narang, S., Gupta, R., & Narang, A. (2011). Probiotics in oral healthcare–A review. *Int J Sci Eng Res*, *2*, 1-5.
39. Nichols, A. W. (2007). Probiotics and athletic performance: a systematic review. *Current sports medicine reports*, *6*(4), 269-273.
40. Novik, G., & Savich, V. (2020). Beneficial microbiota. Probiotics and pharmaceutical products in functional nutrition and medicine. *Microbes and Infection*, *22*(1), 8-18.
41. Parvez, S., Malik, K. A., Ah Kang, S., & Kim, H. Y. (2006). Probiotics and their fermented food products are beneficial for health. *Journal of applied microbiology*, *100*(6), 1171-1185.
42. Pandey, K., Naik, S., & Vakil, B. (2015). Probiotics, prebiotics and synbiotics-a review. *Journal of food science and technology*, *52*(12), 7577-7587.
43. Pratap, K., Taki, A. C., Johnston, E. B., Lopata, A. L., & Kamath, S. D. (2020). A comprehensive review on natural bioactive compounds and probiotics as potential therapeutics in food allergy treatment. *Frontiers in Immunology*, *11*, 996.
44. Prins, A. (2011). Probiotics:" conbiotics" or medicine of the 21st century?. *South African Journal of Clinical Nutrition*, *24*(2), 59-60.

45. Qi, W., Liang, X., Yun, T., & Guo, W. (2019). Growth and survival of microencapsulated probiotics prepared by emulsion and internal gelation. *Journal of food science and technology*, *56*(3), 1398-1404.

46. Reid, G. (2016). Probiotics: definition, scope and mechanisms of action. *Best practice & research Clinical gastroenterology*, *30*(1), 17-25.

47. Rusu, E., Enache, G., Cursaru, R., Alexescu, A., Radu, R., Onila, O., & Radulian, G. (2019). Prebiotics and probiotics in atopic dermatitis. *Experimental and therapeutic medicine*, *18*(2), 926-931.

48. Singh, T. P., & Natraj, B. H. (2021). Next-generation probiotics: a promising approach towards designing personalized medicine. *Critical Reviews in Microbiology*, *47*(4), 479-498.

49. Singh, K., Kallali, B., Kumar, A., & Thaker, V. (2011). Probiotics: A review. *Asian Pacific Journal of Tropical Biomedicine*, *1*(2), S287-S290.

50. Sidhu, G. K., Mantha, S., Murthi, S., Sura, H., Kadaru, P., & Jangra, J. K. (2015). Evaluation of Lactobacillus and Streptococcus mutans by Addition of Probiotics in the form of Curd in the Diet. *Journal of international oral health: JIOH*, *7*(7), 85.

51. Sistek, D., Kelly, R., Wickens, K., Stanley, T., Fitzharris, P., & Crane, J. (2006). Is the effect of probiotics on atopic dermatitis confined to food sensitized children?. *Clinical & Experimental Allergy*, *36*(5), 629-633.

52. Saraf, K., Shashikanth, M. C., Priy, T., Sultana, N., & Chaitanya, N. C. (2010). Probiotics-do they have a role in medicine and dentistry. *J Assoc Physicians India*, *58*, 488-90.

53. Stavropoulou, E., & Bezirtzoglou, E. (2020). Probiotics in medicine: a long debate. *Frontiers in immunology*, *11*, 2192.

54. Smolyansky, J. (2010). Probiotics: a historical perspective. In *Bioactive Foods in Promoting Health* (pp. 43-45). Academic Press.

55. Tadesse, S. (2012). Probiotics, prebiotics and synbiotics as functional food ingredients: production, health benefits and safety. *Journal of Biologically Active Products from Nature*, *2*(3), 124-134.

56. Valero-Cases, E., Cerdá-Bernad, D., Pastor, J. J., & Frutos, M. J. (2020). Non-dairy fermented beverages as potential carriers to ensure probiotics, prebiotics, and bioactive compounds arrive to the gut and their health benefits. *Nutrients*, *12*(6), 1666.

57. Wescombe, P. A., Hale, J. D., Heng, N. C., & Tagg, J. R. (2012). Developing oral probiotics from Streptococcus salivarius. *Future microbiology*, *7*(12), 1355-1371.
58. Yao, M., Xie, J., Du, H., McClements, D. J., Xiao, H., & Li, L. (2020). Progress in microencapsulation of probiotics: A review. *Comprehensive Reviews in Food Science and Food Safety*, *19*(2), 857-874.
59. Zhang, G. Q., Hu, H. J., Liu, C. Y., Zhang, Q., Shakya, S., & Li, Z. Y. (2016). Probiotics for prevention of atopy and food hypersensitivity in early childhood: a PRISMA-compliant systematic review and meta-analysis of randomized controlled trials. *Medicine*, *95*(8).

INDEX